Winning New Business in Construction

Winning New Business in Construction

Terry Gillen

GOWER

© Terry Gillen 2005

Published by
Gower Publishing Limited
Gower House
Croft Road
Aldershot
Hants GU11 3HR
England

Gower Publishing Company
Suite 420
101 Cherry Street
Burlington,
VT 05401-4405
USA

Terry Gillen has asserted his right under the Copyright, Designs and Patents Act 1988 to be identified as the author of this work.

British Library Cataloguing in Publication Data
Gillen, Terry
  Winning new business in construction. - (The leading
  construction series)
  1. Construction industry - Marketing 2. Construction industry
  - Customer services
  I. Title
  624'.0688

  ISBN 0 566 08615 8

Library of Congress Cataloging-in-Publication Data
Gillen, Terry.
  Winning new business in construction / by Terry Gillen.
       p. cm. -- (The leading contruction series)
  ISBN 0-566-08615-8
  1. Construction industry--Marketing. 2. Construction industry--Customer services. I.
Title. II. Series.
  HD9715.A2G53 2004
  624'.068'8--dc22

2004023369

Typeset by IML Typographers, Birkenhead, Merseyside.
Printed and bound in Great Britain by MPG Books Ltd, Bodmin, Cornwall.

Contents

List of Figures and Tables

Figures

Tables

As in many industries, construction industry staff are predominantly technical specialists who, at times, are required to market, sell and negotiate – often with little or no training in winning new business. As sound technical skill is only part of what is needed to win contracts, it is all too easy to lose business to better-trained competitors. With the right knowledge and skills, however, you can win more new business, develop a reputation for excellent service, be seen as thoroughly professional – and add directly to the bottom line.

This book will help you understand what happens in marketing, sales and negotiating situations. It will help you learn how to improve your skills in selling, negotiating, writing proposals and making presentations in support of a tender. Finally, as it is always more profitable to win additional business from existing customers than it is to replace a lost customer, the book will also show you how to retain customers by providing excellent service.

This book is divided into seven chapters, each of which is briefly described below.

Marketing

Chapter 1 shows you some of the key concepts in marketing, explains how you can apply them and provides a range of marketing ideas from which you can select those you feel will be most profitable. The section on customer care also provides you with more business-winning concepts relating to marketing.

Essential selling skills

After explaining how selling skills will benefit you, Chapter 2 shows you how our behaviour in sales situations is often counterproductive. It then shows a simple, yet effective, model that not only makes you better at selling, but also enhances your professional skills.

Essential negotiating skills

As with selling, negotiating does not come naturally to everyone. Not only can the wrong skills prove counterproductive, but having the right skills, besides enabling you to do a better job, also enhances your portfolio of professional skills. Chapter 3 takes you step-by-step through the whole negotiating process.

Advanced selling and negotiating skills

Chapter 4 extends the previous two chapters even further by explaining a range of subtle and effective influencing techniques.

Sales proposals and presentations

Written proposals and presentations are commonplace in construction and, if you want yours to be assessed on something other than price, you need some reliable and tested advice. You will find it in Chapter 5.

Winning new business through customer care

You might be surprised at how much customer care can add to your bottom line (or how a lack of it can wipe out your profits). But how much do you really know about it? Does it form part of your business-winning strategy? Chapter 6 contains some eye-opening information and easy-to-implement ideas.

Partnering

Similarly, businesses that enter into partner arrangements can be surprised how many savings can be made by adopting an open and cooperative approach with others in the 'value chain'. Chapter 7 looks at the principles and the practicalities of partnering.

To help you get the most out of this book, each chapter ends with a list of suggested action points.

Chapter 1

Marketing

In this chapter we are going to look at marketing. You will learn about some key marketing concepts and how they can relate to your activities. You will also be able to select, from a wide range of ideas, those that are relevant and useful for you. We will begin, however, by explaining marketing's place in an overall business.

The place of marketing in business

Most businesses exist for a simple purpose – to grow a small quantity of money into a bigger quantity of money.

Figure 1.1 *The business process*

Let us take a couple of extreme examples. First, someone with construction industry experience decides to start up their own company. They use savings or borrowings (the small quantity of money) to get the business up and running, to promote it, to win contracts, to buy or hire essential equipment, to buy raw materials and to pay for staff or subcontractors with certain skills. These are the *inputs*. The individual will then subject these inputs to various processes during which they are converted into *outputs* (for example, someone's patio or kitchen extension) for which a customer is prepared to pay (the bigger quantity of money).

The second example is both very different and very similar. This is a well-established multi-billion pound global enterprise employing thousands of people worldwide. It uses retained profits, shareholders' funds and borrowings (the small quantity of money) to promote the business, win contracts, to buy or hire essential equipment, to buy raw materials and to pay for staff and subcontractors with certain skills. Again, these are the *inputs*. The people running the company then subject these inputs to various processes during which they are converted into *outputs* (housing estates, apartment buildings, motorways and bridges) for which customers are prepared to pay (the bigger quantity of money).

Although the two enterprises are very different, the principle is exactly the same in both examples. Money is converted into inputs; then, the inputs are subject to various processes to convert them into outputs for which customers are prepared to pay. If customers pay the sole trader more than the cost of both the inputs and the processes, he or she makes a profit. If customers pay the multi-billion pound global enterprise more than the cost of the inputs and the processes, the enterprise makes a profit.

The significance of this point lies in the *processes* by which inputs are converted into outputs. Before the sole trader sets up their business, they are probably already experienced in the construction industry. (After all, it is a bit difficult, and very risky, to set up a business in an area about which you know nothing!) They will be a *technical expert* who very probably enjoys working in construction. (After all, why set up a business in an area you hate?) Likewise, the global enterprise employs many technical experts such as architects, surveyors, site managers and so on, who, hopefully, also enjoy working in construction. Herein lies a potential problem – expertise and enjoyment can combine to cause 'tunnel vision'. It becomes easy for the sole trader, when faced with day-to-day activities, to forget that they are no longer just a construction expert, but also responsible for office management, accounts, marketing, strategy, personnel and so on. The specialist in the global enterprise faces a similar problem. Day-to-day activities, plus the tendency of a large organization to 'compartmentalize' functions, makes it easy to forget the underlying purpose of one's job – *to help convert inputs into profitable outputs*. Whilst the large organization might have specialist teams with nominated responsibility for, say, personnel, a site manager, for example, will perform better if they acknowledge how they can contribute to processes that run alongside their specialism.

People need to understand how what they do contributes to the business's profits.

Marketing is one of the processes that runs alongside your specialism and, by learning about it, you improve your performance and contribute more to the business.

What is marketing?

The process of marketing can be viewed both narrowly and broadly.

The narrow view is that it is a formalized activity, handled by a specialist team within a company, which involves analysing the market and promoting the business to certain parts of it that are congruent with the business's strategy. This narrow view is traditionally geared to big business and requires money and specialist knowledge. It tends to focus on 'markets' (which, incidentally, don't actually buy anything). The narrow view, for example, might result in an advertising campaign aimed at a particular market segment.

Everyone *has a marketing responsibility.*

The broader view is that marketing is any contact that any member of a company might have with anyone outside the company that could favourably affect relations with customers. The broad view is available to businesses of any size, requires values and motivation rather than money and focuses on 'customers' (who *do* actually buy things). The broad view, for example, might result in a surveyor spotting a business opportunity with a client and passing it on to the relevant team.

How to adopt the broad view of marketing

Here is a range of ideas that will help you adopt the broad view of marketing. We will begin with some general ideas and then list specific ideas. While the general ideas will be applicable to everyone, you may find that only a few of the specific ideas are right for you.

Conduct a SWOT analysis

SWOT analyses, which (as you may already know) stands for Strengths, Weaknesses, Opportunities and Threats, have been around for years but, perhaps because they are so familiar, a surprising number of people do not use them.

A SWOT analysis will enable you to consider what you are good at but might not be fully exploiting, what weaknesses you have that might cause problems now or in the future, opportunities you may not be fully exploiting and threats that might soon cause serious problems. Below is a questionnaire to start you thinking. You might find that some of the questions overlap; this is to help ensure that you cover all the important points.

SWOT Questionnaire

Strengths

- What are our strengths?
- What are we good at?
- Are we especially cost-effective, flexible, responsive and so on?
- Is our reputation, name or brand a plus point?
- What do customers compliment us on?
- What do we have a good reputation for?
- What are we particularly proud of?
- What distinguishes us from the competition?
- What characteristics of ours might competitors be envious of or concerned about?
- Would our customers agree with all these answers?

Weaknesses

- What are our weaknesses?
- Do our internal processes help or hinder us?
- How well known are we? What is our reputation?
- Does our size help or hinder us? In what way?
- What customer complaints do we experience?
- What could we do better?
- In what way are we indistinguishable from our competitors?
- What characteristics of ours might competitors be pleased about?

Opportunities

- Is the market changing in a way that we can easily capitalize on?
- In what way is the supply of skilled staff changing in our favour?
- Is technology changing in ways that we can capitalize on?
- Can we use modern technology to become faster or cheaper or to promote our business?
- What new developments are taking place in our market or geographical area that we can capitalize on?
- Are any competitors facing difficulty or pulling out of a market, leaving a vacuum that we can fill?

Threats

- Is the market changing in any way that could cause us problems?
- In what ways is the supply of skilled staff changing to our detriment?

- Is technology changing in any way that would cause us to have difficulty coping with our present skills or would allow our competitors to get ahead of us or would let new competitors enter the market?
- What are our competitors doing that could cause us problems?
- What possible or potential developments in the market or geographical area would cause us problems?
- What changes in the political, economic, taxation or legal areas might cause us problems?

Here are four examples of the way in which a SWOT analysis might help.

1 With stamp duty and other taxes, the cost of moving house in the UK has become so expensive that many people are choosing instead to extend and improve their current homes. This has created a huge *opportunity* for small builders and for construction companies prepared to operate in this market. It is a *threat*, however, for small builders already in that market who choose to rest on their laurels or whose *weakness* is that they cannot rely on subcontractors or whose staff are used to working on building sites and not in customers' homes.

2 The trend in the UK towards privatization and contracting out has provided an *opportunity* for companies who have a *strength* in navigating through formal tendering procedures or who have the expertise to offer facilities management (say, of a housing association's stock).

3 The British population's love of DIY combined with the plethora of property development and house renovation programmes on television has created a *threat* for companies operating in the simple renovation market, but an *opportunity* for those prepared to help buy-to-let property owners get their property on the rental market as quickly as possible.

4 The trend towards downsizing and outsourcing is a *threat* for companies that suffer poor relationships with subcontractors but an *opportunity* for companies with close and reliable working relationships with quality suppliers.

Decide what business you are in

If you have never asked this question, it is a real eye-opener. If you have previously asked the question and accepted the first obvious answer, asking it again will still be an eye-opener.

A real-life example is supplied by Black & Decker which, at one time, believed itself to be in the power tool business. In fact, times once became so bad that they were nearly out of it! After some careful thinking, the company decided that it was in the *gift* business as a result of its discovery that the market segment with the most potential was people buying power tools as presents for someone else. Consequently, the company redesigned its products' packaging and advertising and went from strength to strength. Rolex is another good example. Even though Rolex is known the world over as a manufacturer of watches, the chief executive openly admits that he doesn't know or care what is going on in the watch business – because he is in the *luxury* business.

So, obviously, you are in the building or the construction business. But what business could you be in? The solutions business, the facilitation business, the dreams business, the value for money business, the investment business? The more ideas you come up with the better, especially if you link them to your customers' problems or concerns or link them to the reasons why customers choose you rather than the competition (or vice versa). Whatever answer you come up with, you will still be building or constructing things, but you will find that your marketing becomes much more incisive.

When did you last ask, 'What business are we in?

Create a benefits list

Why should someone do business with you? The answer is because it would benefit them in some way. Technical specialists, however, do not always think in terms of benefits; they think in terms of product *features* – for example, this contract will take a certain number of months, involve a certain number of people, cost a certain amount of money, require certain construction techniques, result in a building of a certain shape and so on. Customers, however, tend to be more interested in *benefits*. For example, this company was quicker than others; which meant I got ahead of schedule; this company was more cost-effective than others, which meant I saved money on my budget; this company itemized everything, which

Apart from price, why should your customers do business with you?

meant I was confident that I knew where the expenditure was going; this company kept me informed throughout, which meant that I was reassured that the activity was on schedule; this company helped me build a better reputation within my own company.

Notice that customer benefits can be tangible (saving money) and intangible (feeling confident). To help you assess what you supply from a customer's perspective, you can construct a benefits list.

Benefits List	
Characteristics of what we do	Which, for customers, means ...

Think about your 'elevator statement'

How long does it take an elevator to go from one floor to the next? Ten seconds? In that time, if someone asked 'What does your company do?', could you tell them in a descriptive, incisive and attractive way? Would you say, 'Oh, we're a construction company.' Or would you say, 'We help local councils provide quality housing at minimal cost.' Or perhaps, 'We help supermarkets get maximum value for money when designing and building new stores.'

Of course, I'm not suggesting that you actually talk like this so why am I asking you to bother thinking about your 'elevator statement'? Quite simply, because we get a lot of practice at thinking of what we *do* and very little practice at thinking about *why* we do it and *what it means* to a customer – even though that is the decisive reason why customers do business with us. Your 'elevator statement' helps you identify what business you are in and identify your USP (unique selling point).

USPs

Your unique selling point can be defined as: *the single decisive reason customers should do business with you rather than your competitors.*

USPs can be tangible or intangible. In terms of a tangible USP, you might be the only company who does what you do in your area. You might have a patented process or a sole distributor agreement that is the envy of your competitors. You might be cheaper or quicker than anyone else. (Note, however, that cost and speed are not always the best USPs because, unless you can capitalize on economies of scale in ways that your competitors cannot, it is relatively easy for a competitor to beat you on these criteria – or just wait until you go out of business.) Criteria relating to intangibles, on the other hand, are much more difficult to emulate.

So what are intangible USPs? You might be able to stand out on strengths such as experience, reliability, consistency, reputation or customer feedback. You can also turn strengths into USPs. If, for example, you are particularly customer-oriented, you can emphasize that. If you take some of the hassle, anxiety or stress away from customers you can emphasize that. A great way of constructing a USP is to follow this line of reasoning:

Some construction companies' clients have difficulty finding a construction company that ...

which, for them, means that...

so if we ...

we'll have a USP.

Identifying customers' problems and swapping them for solutions is a clever way of finding a USP that customers will appreciate.

Think about what your customers need and swap these needs for solutions you can provide.

Swap problems for solutions

In their interaction with construction companies, what do customers find time-consuming, tedious, problematic, confusing, difficult and so on? (If you do not know the answers to these questions, you had better hope that your competitors do not ask them before you do!) How might you be able to alter what you do or how you do it to swap that problem for a solution thereby making your company more attractive to customers?

Use creative emulation

Sometimes, when working in a job, it is possible to get too close to it and 'not see the wood for the trees'. One way in which you can enhance what your company does is by using creative emulation. Ask yourself the following sets of questions:

- What do your competitors do better than you? Can you go one step ahead of them? What simple value-added addition will make your company more attractive than theirs?

- What constitutes best practice? What do the really admired companies do? Even if you are not in their league, how can you adapt and tailor what they do so that your customers appreciate the difference?

- What do successful companies from other industries do that you could adapt and tailor to your company or could stimulate ideas that could be used in your company to set you apart from your competitors?

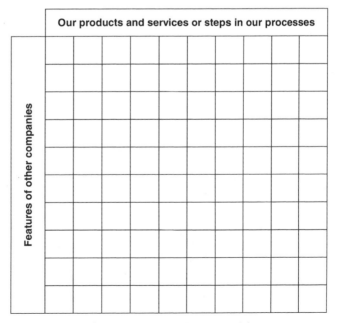

Figure 1.2 *Features matrix*

A simple device called a features matrix can help your creative emulation. Along one axis, list your company's products and services or steps in your processes. Along the other axis describe what strong competitors do, the best practice or the features of successful companies from other industries. Then in the squares created by the matrix, describe how you could adapt and tailor those ideas to your products, services or processes.

Consider new business areas

One way of staying ahead of the competition and winning new business is to expand into new areas or to move from a crowded part of the market to one where you feel there is more opportunity. However, this may take you into unfamiliar areas, and it is in unfamiliar areas that costly mistakes happen. So here is some guidance to help you get it right first time.

Make a list of possible business areas that you could pursue but have not yet done so. Investigate each of them by talking to at least 20 knowledgeable people. To avoid the possibility of people being reluctant to share information with you, be prepared to search for people who will not see you as a competitor. Some will still not want to speak to you but others – often a surprisingly large number – will be happy to share their experience and views with you, especially if you speak to them at a social gathering, meeting of a professional association, conference or such like. Ask them:

- What do you like and dislike about the business?

- What are the biggest problems you encounter?

- What would you do differently if you were starting again in this business?

- What is the biggest lesson you have learned in this business?

- What aspects of the business are more promising than others? Why is that?

This way, you not only avoid numerous pitfalls, but you also avoid the all-too-common phenomenon of considering a new business venture and seeing what you want to see. Remember the old saying: *There might be a gap in the market, but is there a market in the gap?* Getting objective answers is just one benefit of asking the questions listed above.

Let others save you marketing time

It is probably possible to spend so long searching the market for business that you have no time left to actually do any. That is why many construction businesses subscribe to services that search for new business for them.

One such service is offered by Emap Glenigan which has been tracking the construction and property markets since 1973. It reports on every construction or civil engineering project in the UK, tracking development from the early planning stages through the tendering stages to the award of contracts. It also researches the commercial property market, reporting on companies that are moving, expanding and improving premises. The information is updated daily and is available by subscription. Another company that can save you time is ABI which searches every planning application made in the UK, providing customers with up-to-date information on 10 000 projects each week, verifying them and tracking them from tender, through contract to subcontract stages. Not only will both companies save you time in uncovering potential business, they will also save you even more time by providing only the information that suits you thereby relieving you of the need to wade through reams of irrelevant information. Their websites (www.glenigan.com and www.abibuildingdata.com) contain far more details than those provided above and will give you lots of ideas.

Save time by using all the sources of potential new business, that are available.

You can also identify architects and surveyors with whom you might be able to establish mutually beneficial relationships by performing a tailored search on the websites of their respective professional bodies (www.riba.org and www.rics.org).

Check your thinking

Which type of thinking do you believe results in sustained business success?

How can I get customers to pay more without increasing the value I give them? How can I pay my suppliers less while squeezing more value from them? How can I get the best deal for me in negotiations regardless of what the other party(ies) might lose? How can I get my staff to work harder and longer for the same or less money?

Or:

> *How can I help my customers be more successful? How can I help my suppliers be more successful or get more out of their relationship with me? How can I help them prosper? How can I help my staff feel excited about their work and feel glad that they work for me?*

The first type of thinking is very self-centred and selfish. Typically, it results in short-term gain and long-term loss. The second type of thinking helps ensure good longstanding relationships with the three categories of people you cannot do without – your customers, your staff and your suppliers.

Marketing ideas

Here is a list of marketing ideas, many of which cost little, if any, money. Which ones can you use to market your company?

1 Plan your marketing so that it contributes to your business plan and business cycle. Concerted and mutually supportive activity always produces better results than random uncoordinated activities.

2 Plan your marketing on a rolling 12-month cycle. This way, it is always 'live' and flexible and facilitates financial decisions.

3 Ensure that your company name is easy to spell and pronounce, is descriptive of what you do (especially your USP), that it does not prohibit future expansion and under no circumstances reminds customers of your competitors. (There are two major car windscreen replacement companies in the UK with very similar names. When the bigger company advertises, the smaller one always experiences an increase in business!)

4 Use a logo that helps you stand out from the crowd and, if possible, one that illustrates what you do or what you stand for. The vast majority of people are visual thinkers (that's why road signs are images and not sentences), so a good logo is a quick route into their memories.

5 Have a clear, illustrative and truthful strap line. A strap line is a set of words that summarizes what you stand for or offer. Avis, the car rental company, has used the 'We

try harder' strap line for decades. It's simple and effective. Some parcel delivery companies 'never forget that we deliver your promise'.

6   Ensure that your letters, e-mails, proposals, business cards, gifts, promotional literature, websites, vans and trucks and so on carry the same logo and strap line. It 'drip feeds' customers' subconscious minds.

7   Ensure that your customers can access you with sales enquires when it is convenient to them. The more global we become, and the more we work 24/7, the more important easy access becomes.

How easy is it for customers to reach you?

8   Seek referrals from satisfied customers. They may be able to put you in touch with a prospective customer and, as you come with a trusted recommendation, you will not be cold calling.

9   Raise your profile in your 'community'. This could be a geographical community or a professional community. Be seen in the relevant journals and newspapers. Write letters to the editors responding to articles or other people's letters, write articles or columns that establish your company as a leader, issue press releases, let journalists know that you are available to comment on 'community' issues.

Whatever your company size, raising your profile can make a huge difference.

10   Obtain copies of articles referring to your company and send them to existing and prospective customers.

11   Showcase successes in your premises, in your brochures, on your website and so on.

12   Ensure that your website is regularly reviewed. It needs to be designed for ease of use, quick download, relevant information and maximum 'stickability'. A website is often the first and prime information source for prospective customers gathering information about you.

13   Ensure that key people actively network. When a prospective customer needs a company like yours, it is amazing how many people like to recommend 'someone who might be able to help'.

14   Adopt a cause, such as a charity. The charity benefits and so does your company's image. Also, staff tend to enjoy raising funds for good causes.

15   Become an educator. Provide free seminars or speak at conferences (many of the attendees will be prospective

customers). Make free factsheets available. Many of the people who request them will be prospective customers and, if you offer them through your website or a faxback service, you minimize administration.

16   Conduct some relevant research attractive to prospective customers, publicize it (through press releases, articles, radio interviews and so on) and make it freely available.

17   Advertise jointly with complementary businesses to reduce costs.

18   Form relationships with other companies that might pass business leads on to you and do the same for them.

Other businesses will gladly recommend you to clients who ask for help.

19   Visit your own website and telephone your own company and experience what customers experience. Ask yourself, 'Is this the experience I want my customers to have?'

20   Train all staff who have contact with customers in customer care attitudes and skills. If you use sub-contrators, insist that they are similarly trained. Better still, establish such long-term relationships that you can train them yourself. After all, people don't have to be employed by you to lose or win you future business.

21   Follow up customers after a contract ends to check that they are still happy. If they aren't, you have an opportunity to do something about it before they tell their family, friends, neighbours and colleagues.

22   If you advertise in newspapers or journals opt for repeated smaller adverts than one big expensive one. It's the repetition that gives advertisements their effectiveness. If you advertise in magazines and journals, choose carefully. Is the magazine or journal read by your prospective customers? It may have a huge circulation, but is it a relevant circulation? Will advertising in it add to, or reduce, your company's credibility?

23   Obtain testimonials and incorporate them into all advertising and proposals.

24   Mail existing and prospective customers with something interesting or useful to them, not just your advertising. Ensure that all mailshots are 75 per cent useful to customers and only 25 per cent advertising.

25   Remember that all marketing contains two messages –
the conscious one and the subconscious one. The
conscious one is the stated message 'We sell x'.
The subconscious message *creates feelings* in customers.
As a result of your marketing, do they feel that you are
superficial, too slick and unimaginative or professional,
good value, reliable, reassuring and so on?

Action points

1   Ask yourself why you do this job. If you own or part-own
the company, it should be because you enjoy it and
because, one day, you want to sell it. If you are an
employee it should be because you find it a satisfying
way of earning a living. If this sounds strange, ask
yourself how enthusiastically you would market the
company because, if you can't do this with enthusiasm,
the chances are that you're simply doing a job, not
helping convert the inputs into profitable outputs.

2   Conduct a SWOT analysis with several people. Make it a
team event, preferably off-site. Perhaps consider inviting
customers who you get on well with and who will give
you honest answers. If you want it to be really thorough,
show the analysis to other people inside and outside your
company, such as some friendly competitors, your bank
manager or an academic who knows your market, and
seek their input.

3   Begin an off-site teambuilding event with the question
'What business are we in?' or 'What is our elevator
statement?'. Encourage creative answers and give a prize
for the one that everyone feels is the most incisive and
exciting. See if you can turn it into business strategy.

4   Get the team to create a benefits list in the same way.

5   When a new manager joins, give them a week to create a
SWOT analysis or a benefits list or a list of USPs by
visiting anyone and everyone who might have something
useful to say. It will be a great induction for them and an
eye-opener for you.

6   Get staff to tell you what problems their work has solved
for customers. This will encourage them to think
'problem-solving' and will provide useful marketing
information.

7   Set up a project team to review all lost proposals from the previous 12 months, to find out why and to suggest improvements.

8   Set up a project team with the brief to find and evaluate new business areas, including the use of directory services as sources of new business.

9   Charge all your senior people with doing one thing in the next three months (such as writing articles or speaking at a conference) that promotes your business.

10   Review all 25 marketing ideas and implement at least five.

Chapter 2

Essential Selling Skills

In this chapter we are going to look at the sorts of sales skills that will complement and enhance your professional skills – the sort that make it easy for customers to accept your advice, suggestions and proposals. They are also the sorts of skills that will distinguish you from your competitors.

Professional selling

Broadly speaking, there are two categories of salesperson. If you met one category – say, at a party – and asked them what they did, they would tell you that they were in sales. This year they might be selling motor cars; next year they might be selling timeshares. If you met the other category and asked them what they did, they would tell you that they were an architect, a lawyer, a scientist, a surveyor or whatever. The first category sees themselves as sales specialists first and foremost. The second category sees themselves as specialists in their trade or profession and may not even realize that selling is an important part of their job, not only for their own performance reviews but also for the success with which the business processes inputs into profitable outputs.

Traditionally, selling has relied on techniques such as powerful openings, distinguishing between features and benefits, spotting buying signals, overcoming objections and creating irresistible closes. This is fine as far as it goes (and some of the techniques are useful) but, to ensure that you behave as a professional specialist rather than a salesperson, you need to go further because you need to be, and be perceived as being, a credible professional who builds productive long-term relationships with customers.

The right 'sales skills' make you even more professional.

But, as a credible professional, why do you need selling skills? Quite simply, if you need customers to go along with your ideas, to accept your recommendations, to select your proposal rather than a competitor's or to rethink a hasty decision, you need the skills of persuading and influencing or, as we usually say, 'selling skills'.

This chapter, based on a practical tried and tested model, explains the mental processes customers go through during a sales situation and will show you how untrained people, no matter how good they are technically, can inadvertently trigger a negative response. It will then show you how to encourage a positive response from customers.

The information contained in this chapter will not only help you when selling, it will also help you in any situation where you are trying to influence someone, such as in meetings or when you are discussing contracts with a customer's specialist, such as an architect or surveyor.

We will begin with some fundamental information about what happens when we try to influence someone.

Whose reasoning?

Whenever we try to influence someone there is a simple trap we tumble into time and time again: we assume that the reasoning that makes sense to us will also make sense to them.

Does what you say make sense to your customers or just to you?

You can tell when you have fallen into this trap because you get the feeling that the other person is not listening, does not care, is being obstinate, is being selfish or is just plain stupid. Sometimes, of course, you are right! Sometimes, however, these thoughts are just ways of shifting responsibility for a successful outcome away from ourselves because we are the ones who have actually triggered these responses! Using the reasoning that makes sense to us instead of the reasoning that makes sense to them is an effective way of triggering resistance. You will appreciate this point more easily, and understand what to do about it, if you first understand *the triggering process*.

The triggering process

Behaviour from one person will trigger a response from someone else. If, for example, you present a proposal as a statement

We must delay the start date

other people's natural and almost automatic reaction is to resist, disagree, find fault and so on. Because your statement is presented in a way that excludes them and even orders them to do what you want, and because they are not privy to your thinking, it is natural for them to react like this. If, on the other hand, you present your proposal as a suggestion

I'm wondering if we could consider delaying the start date

you will be much less likely to trigger resistance because the presentation is less dictatorial and more invitational. Furthermore, if you give a reason and express a concern

If the other contractors run over, as you fear they might, I'm concerned that you'll be paying me unnecessarily so I'm wondering if we could consider delaying the start date

Talk the language of your customers. Don't expect them to talk your language.

the customer understands you, hears what you say as reasonable (and even in their interests) and is much more likely to agree with you.

The key point in this example is that behaviour from one person triggers a response from the other person.

Trigger   Response

As our interpersonal skills are often 'tested' when trying to influence someone, we can easily and inadvertently use behaviours that are destined to trigger a negative, rather than a positive, response from the other person. How many times have you used any of the following 'foot-shooting' behaviours?

- Trying to get them to 'see sense' by impatiently explaining your point again. You may not mean to sound impatient or even realize that you come across as impatient, but the other person will notice it and react negatively.

- Using more insistent language such as telling them what they *should*, *must*, *ought* or *can't* do. The more people feel they are being pushed into a course of action, the more they naturally resist.

- Using phrases such as 'I hear what you say but…', 'Let's be realistic', 'Let's be honest', and 'Look, I'm being perfectly reasonable'. On hearing these phrases the other person assumes that you are accusing them of being unrealistic, dishonest and unreasonable, and reacts accordingly.

- Using phrases such as 'What you don't seem to realize is…' or 'What you have clearly forgotten is…'. These phrases make us come across as impatient and condescending.

- Digging in your heels (or, by your tone of voice, nature of eye contact and posture, seeming to dig in your heels) and insisting. All this behaviour often achieves is to encourage the other person to respond in kind or to get their revenge later.

- Disagreeing with their reasons 'I disagree with you because…' or telling them that their idea 'just won't work'. These trigger phrases usually provoke counter-disagreement and a more forceful justification of the other person's position.

- Listening to their proposal and then, without discussing it, coming up with an even better one. Even if your proposal is excellent, mentioning it when the other person is still wondering why you have just ignored their perfectly good (to them) idea is guaranteed to trigger a negative response.

- Signalling that you are expecting their refusal by saying, 'I don't suppose it would be possible to…?' or 'You probably won't be able to agree to this but…'. This will probably invite exactly the response you do not want.

The 'trick' therefore is to remember that your behaviour triggers a response from the other person and therefore to use behaviours that dramatically increase the chances of a positive response. Several of those behaviours combine to form a framework called the 'persuasive funnel'.

Your behaviour is like your other tools; you need to use it skilfully.

The persuasive funnel

When we are trying to persuade someone, our case is probably a sound one – to us. It is all too easy, however, to assume that they and we share the same reasoning. That can tempt us into trying to persuade them by using the reasoning *that makes sense to us*. This approach is rarely successful. To gain their commitment, the penny needs to drop in *their* mind. This means that it is more likely to drop if we use their reasoning. But how do you find out what their reasoning is? The answer is simple: you ask them.

The only trouble is they may not know all of it themselves! Human beings are complex animals. Our thinking is made up of *rational thoughts* and *emotional thoughts*. Sometimes the emotional ones are not always obvious – even to ourselves. So simply asking the person may not be enough. We need to be good at *exploring* their thinking but exploring with an end in mind – gaining their commitment.

Consequently, we need a framework that ties together an understanding of their reasoning with our aim of gaining their agreement. This framework is known as the *persuasive funnel*.

The persuasive funnel has three stages. In the first stage you explore their reasoning by asking questions and probing into their answers. Your aim is to understand what they want, why they want it, how they see things working in practice, what their priorities, concerns and anxieties might be and so on.

In the second stage, you summarize what they have said. This lets them know you have been listening to them and that you understand them. It also brings to the fore key points that, when emphasized, lead naturally to the third stage.

In this last stage, you make your recommendation or proposal, but you do so as a suggestion rather than an 'order'

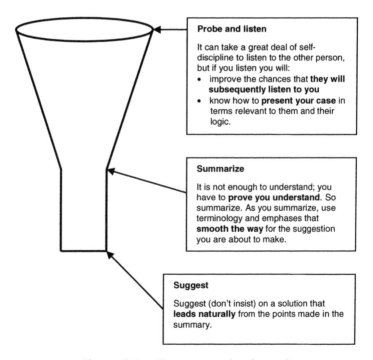

Figure 2.1 *The persuasive funnel*

because, as described above, suggestions are more inviting, more thought-provoking and more involving, all of which increases the likelihood that they will agree with you.

The following two example dialogues illustrate how the persuasive funnel works in practice. The conversation is an initial discussion between a specialist from a construction company (S) and a potential customer (C) to determine whether to include the construction company on a tender list. In the first dialogue the construction company specialist does not use the persuasive funnel and in the second dialogue he does.

Example 1

S: Thank you for seeing me, Mr Smith. As I mentioned over the 'phone, my company is a specialist in property renovation and maintenance, and I'm sure we can offer you a really good deal.

C: That's as maybe, but I already have several potential suppliers shortlisted. I don't really want to add to that list right now.

S: If I were you, I'd think carefully; because we don't use subcontractors we can rely on all our personnel.

C: I understand that, but I don't have the time to verify everything you say, especially as I've got several companies shortlisted already.

S: But, surely, if we can do a better job …

C: I'm not saying you can't do a better job. I'm just saying that I've already got several companies shortlisted and I don't have the time to put you through our selection procedure to put you on the shortlist as well.

S: But isn't reliability important to you?

C: Of course it is, but so is my time. I've already got a full in-tray sorting out tenants' problems with the new bathrooms we installed last year so, if you'll excuse me, I have to get on.

Example 2

S: Thank you for seeing me, Mr Smith. As I mentioned over the 'phone, my company is a specialist in property renovation and maintenance. I recall, however, that you mentioned that you already had enough companies on your shortlist. You also said you were short of time, so let me be brief. How do you determine whether to put a company on your shortlist?

C: I check them to make sure they've got the experience and resources to do a good job.

S: How do you do that?

C: I look at how long they've been in business, who else they've worked for, how many people they employ, which subcontractors they use. That sort of thing. It's the subcontractors that usually present the problems.

S: In what way?

C: Most subcontractors are just after a quick buck. Who can blame them? If their work goes wrong months later they're not around to sort it out, so another subcontractor is called in who might be as bad as the first one. OK, the company doesn't have to use really bad subcontractors again and they all say they can rely on the subcontractors but, in reality, good subcontractors go where the money's best and managers, being under pressure to keep costs down, go for the cheapest ones.

S: You mentioned experience as well. Which is more important to you, the experience or the resources?

C: At one time I'd have said the experience but, since the problems with the new bathrooms, I'd say the resources are more important.

S: What happened with the new bathrooms?

C: We have big renovation plans for our housing stock and six months ago we began to upgrade the bathrooms on one of our estates. Our current contractor put the job out to tender and used a series of subcontractors. I'm now getting far too many complaints from tenants about the standard of work. My contractor is addressing them but also arguing about some of them. Now it's got down to arguing about the small print in the contract. Quite frankly, it's hassle I can do without. We've got four more estates to renovate so now I'm going through the shortlisted companies to see who I'll invite to tender to renovate the remaining estates.

S: Let me make sure I understand you. You feel too busy to review more companies to see if they could go on your shortlist but, at the moment, most of your time seems to be taken up with problems caused by your current supplier's use of unreliable subcontractors.

C: Yes, that's right.

S: And how confident do you feel that the other companies on your shortlist will be any better?

C: Hmm. That's a fair point. They all use subcontractors. Probably the same ones for all I know.

S: How much more confident would you feel if you were dealing with a company that had a solid core of its own skilled staff and took over the job of liaising with tenants via its own call centre?

C: Tell me more. This sounds interesting.

S: I'd be glad to.

In the first conversation, the construction company specialist did not probe into anything the customer said which gave the impression that he was not really listening. Consequently the points he made sounded as if he was coming up with a counterpoint or even disagreeing, which triggered the customer's negative response.

In the second conversation, the construction company specialist probed into what the customer said, identified key points which he then summarized and which, in turn, led naturally to the suggestion of how he could help.

If you ask your customers what they need, they usually tell you.

Using the persuasive funnel gives you four important benefits:

1 The other person is more likely to listen to you because you've just listened to them. We sometimes feel that, to make our point, we need to get in first. Unfortunately, speaking first has two disadvantages. You may not know how to make your point so that it appeals to their reasoning, and they are so busy thinking of their points that they may not really be listening to you.

2 You know how to tailor what you say to fit their reasoning. This is a major advantage in letting them speak first. They tell you what is important to them and what their concerns are.

3   While they are talking you have time to think. People speak at about 165 words per minute, but our brains process information at the equivalent of 2–3000 words per minute. (Not only do we take lots of shortcuts in our thinking, some of it is intuitive, which is very fast indeed.) When you are listening, therefore, you have spare thinking capacity which you can use to marshal your summary and suggestion. If, however, you try to talk and think simultaneously, you will probably fumble over your words, make mistakes and come across as unprofessional.

4   Letting them go first, listening to them and tailoring what you say to their reasoning helps them feel better about the discussion, the decision and you. You come across as a knowledgeable, customer-focused professional who understands their needs.

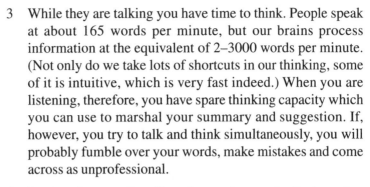

Customers usually do business with suppliers with whom they feel comfortable.

Exploring someone's thinking sounds easy, but it isn't. We all have four in-built obstacles. These obstacles are perfectly natural to us, which means that everyone is prone to them. As you read them, you might like to consider how easily they can give rise to the 'foot-shooting' behaviours described earlier.

Obstacle	'Foot-shooting' response
1 Our different perspective. We see things from our perspective more easily than we see things from someone else's perspective.	We try to influence them using the logic that makes sense to us instead of the logic that makes sense to them. So, even when we try to probe, we ask questions like *'But don't you agree with me that ...?'* and come across as manipulative.
2 The differential between their speaking speed and our thinking speed. While they are speaking at about 165 words a minute, we are thinking at the equivalent of 2–3000 words a minute.	We get ahead of them in the conversation and ask questions that *check* what we are thinking rather than *explore* what they are thinking. So *'Why do you think that way?'* becomes *'Why do you think that way. I mean, is it because of ...?'* or *'But don't you think that ... [your point]?'*
3 Our brain's natural function to recognize something familiar in what it receives and send a transmission straight to our vocal cords.	Here are some illustrative examples: Them: *'We were on our way to Disneyworld when this guy comes out of a side street and collides with us completely wrecking the car and ...'*

continued

Obstacle	'Foot-shooting' response
	You (interrupting): 'We went to Disneyworld last year. Isn't it great?' You were hearing (otherwise you would not have heard the word 'Disneyworld' but you were not really listening which is why you did not register their key point – the car crash. Them: 'So I feel strongly that what we need to do is [their idea].' You (because their idea has just triggered another idea in your mind: 'Or we could [your idea]. By apparently dismissing their idea they will probably reciprocate by dismissing your idea.
4 Our **brain's natural function to evaluate** information it receives.	Them: 'So I feel strongly that what we need to do is [their idea].' You: 'That won't work.' Or 'Yes, but what you're forgetting is ...' You might be right, but expect them to disagree with your disagreement or to stubbornly justify their idea.

concluded

Table 2.1   *Obstacles and 'foot-shooting' responses*

Action points

1   Observe others in influencing situations and see how many of the 'foot-shooting' behaviours you can spot. Notice the effect they have on others.

2   Observe others in complex (professional) selling situations and see how they attempt to sell. Note how many do it by talking and how many do it by probing and listening. Notice the effect on the customers.

3   Reflect on situations where you were trying to sell to a customer and try to calculate the ratio of talking in the early part of the conversation. (This is the wide part of the funnel where the ratio should typically be well in the customer's favour as they respond to your probing.)

4   As you leave a sales situation, find a quiet moment and write down the customer's reasoning. (If you can't do that,

ask yourself whether you were trying to persuade them using your reasoning.)

5 Take your team for an off-site 'development day' and include one exercise where they have to persuade each other of something. Then explain the persuasive funnel to them and get them to do three more such exercises. At the end of each exercise record the ratio of talking between influencer and influencee (it should be well in the influencee's favour by the last exercise). Discuss and agree how they can transfer the persuasive funnel to their work with customers.

Whose reasoning will a customer use to make a purchase decision – yours or theirs?

6 Ensure that all staff engaged in selling situations receive quality training in selling or influencing skills.

Chapter **3**

Essential Negotiating Skills

In many situations, selling and negotiating will overlap considerably. You will understand negotiating more easily, however, if I explain it separately.

This chapter will explain key concepts in negotiating such as linking, trading, best possible outcome, fallback position, Batna (Best Alternative to a Negotiated Agreement) and so on. It will also show you a variety of ways in which you can improve your preparation, from subtle 'seed sowing' to the pragmatic 'thinking hat' preparation.

Finally, to be as practically helpful as possible, this chapter will also cover the manipulative tactics some negotiators use (which many readers will recognize) and explain the best way of responding to them.

The negotiating process

Effective, day-to-day negotiating is not about ploys and gambits; it is about the right attitude and sound interpersonal skills linked to a helpful framework or process.

The following is an overview of the negotiating process. You might need all of it or only some of it.

- **Preparation**
  This section covers bargaining power, information-gathering, understanding yourself and the other party (or parties), thinking about the outcomes you want to achieve and, finally, the concept of 'power' and the importance of early 'seed sowing'.

- **Exploring**
  This section covers in-depth probing or exploring and understanding the other party's wants, desires and motivations, distinguishing between wants and needs and understanding motivation and how it applies to negotiation.

- **Linking**
  This section shows you how to increase your bargaining power by being more flexible, thinking more laterally and by understanding the difference between costs and benefits.

- **Trading/influencing/persuading**
  This section provides a range of essential ideas to help you negotiate, influence and persuade. It concludes by showing you how to spot and control the manipulative tactics that some negotiators use.

- **Implementing**
  This section shows you the points to watch to ensure the satisfactory implementation of your negotiating agreement.

- **Continual learning**
  This section shows you how to continue developing your negotiating skills and provides useful ideas for developing your negotiating as a life skill.

Preparation

Bargaining power

Negotiating is, in essence, about *trading*. Two children who collect such things as footballer cards or Pokémon cards will know that, while some cards are common, others are comparatively rare and that the rare ones are worth more than the common ones. Consequently, they will trade a number of common cards for one rare one. The exact quantity of cards will depend on their relative bargaining power. If you are the only child in the school with that rare card, your bargaining power is high. If a few other children also have that rare card, your bargaining power is not so high. You will trade better if you have more bargaining power.

Bargaining power can be real or perceived. The bank robber who holds a gun to a hostage's head in the movies has more bargaining power if the police do not realize that the gun is a toy. The poker player who can keep a straight face has more bargaining power than the one whose face displays his every emotion.

In short, because bargaining power is so important when negotiating, one of the main reasons for thorough preparation is to enhance your bargaining power. Although, in business, an ability to bluff is not the best way of winning business and building long-term relations with customers, it is worth being

Negotiating is too important to 'wing it'; proper preparation is vital!

aware that bargaining power is based on perception. Understanding it will enable you to avoid unintentionally eroding your own bargaining power and contributing to that of the other party.

Bargaining power can be affected by a range of factors. Think of a recent or a typical negotiation and ask yourself which of the following factors affected you positively and which affected you negatively:

- Expertise and knowledge

- Credibility

- Reputation

- Expectations

- Confidence

- Personal relationships

- Feelings of obligation or belligerence

- Perceived 'clout'

- Relative status

- Age and experience

- Formality/informality

- Location

- Other priorities and pressures

- Timing.

When you consider everything that can affect your bargaining power, you can see that thorough preparation is vital. Here are some ideas to help you prepare.

Information-gathering

Knowledge improves bargaining power!

In most negotiations, whether formal or informal, knowledge-able people have an advantage. This will not only improve your confidence, but will also improve your persuasiveness as well as helping to protect you against manipulation.

So what information would be useful to you? Information regarding:

- the situation?

- the background to the issue?

- facts and figures?

- what has happened before in similar situations?

- people and their personalities?

- people and their priorities?

Knowing what questions to ask is one thing, but you also need to know who to ask or where to get the information. Consider the following:

- Who knows more about this situation than you?

- Who has had dealings with these people before?

- What were the previous outcomes?

- What affected those outcomes?

- What files, databases or websites can you check for relevant information?

Understanding yourself

A vital part of preparation is to understand yourself. It is a good idea to look behind your *negotiating position* (your wants) and think about the *needs* and *desires* that lie behind them. The difference between the two is illustrated by the Israeli–Egyptian 1977 negotiations. Both parties wanted occupancy of the Sinai desert and neither appeared prepared to give up that position. However, when their interests were explored, it turned out that Egypt desired sovereignty and Israel needed security. In the end they agreed that Egypt would retain sovereignty and rule Sinai, but with a demilitarized zone along the border satisfying Israel's need for security.

In the persuasive funnel example dialogue set out in Chapter 2, the construction company specialist *wanted* to get on the shortlist but *needed* the housing association manager to regard him as an alternative to his current supplier.

In any negotiation, being clear about the distinction between your wants and needs will enable you to stay flexible as well as focused on the right thing and more likely to spot creative solutions.

Understanding the other party

Consider the other party's viewpoint. What do they want? What needs and desires underlie those wants? Again, in the persuasive funnel example dialogue mentioned above, the housing association manager *wanted* to put prospective suppliers through his selection process but *needed* reassurance that a supplier would do the right job and not cause him unnecessary problems.

One way of identifying underlying needs is to listen carefully to the questions customers ask and the points they make and, as you do so, to listen for the *underlying* question or point. (This is another reason for using the persuasive funnel framework.) Below are some examples (the italicized statements represent the underlying message):

'We have a process for deciding who to invite to tender.'
'I'll need a good reason for short-circuiting that process.'

'You're too expensive.'
'I need proof that you're worth the extra money.'

'I can see the sense in what you're saying, but I'm not sure my boss will go along with it.'
'Tell me what to say to my boss so that I can get them to agree.'

Understanding the underlying as points gives you much more room for manoeuvre as well as improving your preparation.

'Thinking hat' preparation

When preparing for negotiations, you do not just need to put on your thinking hat, you need *different types* of thinking hat. Why different thinking hats? Because you need different types of thinking. You will need to analyse the situation from your perspective, from the other party's perspective and perhaps from the perspectives of other interested parties. You will also need to find faults and weaknesses in your case before the other party finds them. Have you ever considered how much you might benefit from asking someone to play 'devil's advocate' and try to undermine your case as if they were the other party or perhaps from asking them to find strengths in the other party's case.

When you adopt 'thinking hat' preparation, you can do so in a productive sequence:

1   What are the relevant facts and objective data?

2   What are my wants and needs?

3   How will the other party feel about them?

4   What are the other party's wants and needs and how can I address them?

5   What are the potential weaknesses in my position and the potential strengths in the other party's?

6   What can I do to improve my bargaining power, both real and perceived?

Thinking about outcomes

Another vital aspect of preparation is to think about outcomes.

Inexperienced negotiators can be so convinced by their own logic that they sometimes ignore the many factors affecting negotiations and, in so doing, develop an unrealistically optimistic expectation of the outcome they can achieve. Alternatively, they can be so overwhelmed by factors such as the other party's status that they develop an unrealistically pessimistic expectation of the outcome that can be achieved. Hence, it is a good idea to think *objectively* about:

Proper preparation can compensate for limited experience.

- a perfect outcome – what could be achieved if you aimed really high

- a likely outcome – what could be achieved if the negotiation proceeds smoothly

- a fallback position – the minimum you will accept or the maximum you will concede if the negotiations go badly

- a Best Alternative to a Negotiated Agreement (Batna) – what you could do if negotiations fail.

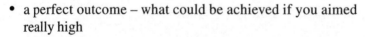

This kind of preparation aids both planning and confidence. If, for example, you are negotiating alongside a colleague, it saves you the unfortunate consequences of one of you conceding a point, just as the other was about to stick on the same point.

Sometimes the Batna is so surprisingly attractive that you can confidently raise your fallback point. For example, failing to win a contract for rail maintenance may be no bad thing if the insurance costs, in the wake of recent rail accidents, are about to go sky-high. Alternatively, sometimes your Batna can be so unattractive that you realize that even an outcome in which you concede a great deal is preferable to one in which you fail to reach agreement at all. For example, if one of your principal customers has gone bust, leaving your invoices unpaid, you may need this contract even if it only covers costs.

The way to steer clear of an unattractive Batna or low fallback point is to improve your bargaining power, and one of the easiest ways to do that is by 'seed-sowing'.

The concept of 'power' and the importance of 'seed-sowing'

In the same way that beauty exists in the mind of the beholder, so can power. You can influence the other party's perceptions of the strength of your case in a variety of ways.

Think of a forthcoming or typical negotiating situation and, as you read through the following list, highlight anything that could help you increase your bargaining power.

- Provide information, facts, statistics and so on that support your position.

- Obtain the support of influential people or groups.

- Establish your credibility and expertise in the subject.

- Select a venue that subtly supports your goals.

- Opt for durations and deadlines that support your goals.

- Let the other party see that your approach is based on a problem-solving, rather than a tug-of-war, mentality.

- Let the other party see that attempts to exert power related to status, ploys and gambits and so on do not seem to affect you.

A positive mind-set

Successful negotiation requires realism mixed with optimism and confidence. It also requires a realization that you need this customer to feel satisfied about working with you. Achieving a good deal during negotiations is only a hollow victory if the customer subsequently feels aggrieved, never works with you again and tells other potential customers about their unsatisfactory experiences with you. Read the following list and highlight any attitudes you could usefully improve.

- See negotiating as a joint problem-solving activity rather than as a tug-of-war.

- Be confident enough to be unintimidated by tough negotiators, negotiators who pull rank, manipulative tactics and so on.

- Believe that negotiating is normal rather than something to feel uncomfortable about.

- Believe that, in many situations, successful negotiating can produce an outcome better for both parties than they could reach on their own.

Exploring and information-gathering

In negotiating, information increases understanding and that means more bargaining power. Some information, however, cannot be obtained during preparation; it can only be obtained during discussion – which means you have to be good at in-depth probing.

Probing is an essential skill if you are to uncover the other party's wants and needs, especially where they themselves are not fully aware of them. This is another situation in which the persuasive funnel proves invaluable.

Although using the funnel enables you to avoid the obstacles mentioned earlier, it does require a lot of patience. Genuinely listening to the other person's arguments, assumptions, beliefs, points of view, prejudices, biases and stupidities requires self-discipline. It is worth it, however, because it gives you a head-start over less patient negotiators.

Negotiation requires patient listening as well as preparation.

One of the things to listen for is what motivates the other party.

Motivation

The other party will agree with you if (a) what you say makes sense to them and (b) they are motivated to do so. However, (a) on its own is rarely enough: you need both.

As a general rule, people are motivated by what makes sense to them rather than what makes sense to you. That was why, in Chapter 2, we placed heavy emphasis on the persuasive funnel and, in this chapter, on distinguishing between wants and needs. Effective negotiators need to take another factor into account, however – namely, what motivates the other party.

A motivator is something that causes us to take action. Some motivators are internal and some are external. If, for example, you work hard to achieve a financial bonus, that is an external motivator. If you work hard because you enjoy what you are doing, that is an internal motivator.

Motivators can be many and varied. To prompt your thinking about what can motivate people during negotiations and how you can use that information, Table 3.1 offers some information on the motivators you might come across, how you can spot them and what you would need to do as a result.

Linking and trading

So, having prepared and explored you can begin to link together, in your own mind, things you can trade.

Trading is at the heart of negotiating. Put simply, it means swapping something you have (that the other party wants) for something you want (that the other party has) – 'I'll give you this if you give me that'.

The ideal situation, of benefit to both parties, is where they have something which is 'valuable' to you but which will cost them little and you have something 'valuable' to them but which will cost you little. This way, you both receive maximum benefit for little cost. Real life, however, is rarely as convenient

Motivator	What to look for	What to do
Safety, security Likes to minimize risk and keep things safe.	• Displays of anxiety. • Pointing out how your proposal is different from what they envisaged or currently do.	• Explain how your proposal has built-in safeguards. • Emphasize the risk in alternatives. • Emphasize similarities with the current situation.
Away from Choosing course of action to avoid something.	• A focus on problems and difficulties.	• Explain how your proposal helps remove or avoid problems and difficulties.
Towards Choosing course of action to achieve something.	• A focus on goals.	• Explain how your proposal helps them achieve those goals.
Self-interest Putting their own needs and ambitions first.	• An excessive concern with status. • Tell-tale signs of career aspirations.	• Emphasize how your proposal will show them in a good light, benefit them and so on.
External Heavily influenced by outside pressures; concerned with what others think.	• Frequent references to what other parties (for example, their boss) might think.	• Probe to uncover more information about the other parties and help them 'rehearse' how they will present your proposal to them.

Table 3.1 *Motivators and how to use them to your advantage*

as that, so here are some ideas that will help you link and trade more advantageously.

Develop a trading attitude

Have you ever gone into a shop (by the way, this only works in a shop run by a proprietor) and been about to buy something and asked, 'How much per item if I buy two?' or 'How much if I pay by cash instead of credit card?'. Often the proprietor wants to sell more than one item and would rather receive cash than pay the credit card company's charges and they will respond by giving you a discount. Have you ever had a hotel lose your booking and only have a small room available, and suggested to the manager that a complementary bottle of wine with dinner might restore your faith in them? Often the manager wants to ensure repeat business and regards a bottle of house wine as a tiny investment to ensure your goodwill.

Are you a natural negotiator?

If you have something which customers value but which is easy for you to concede, do you still trade it? They might want a quick delivery, and your delivery truck might be in their area tomorrow as a matter of routine – but do you still *trade* it? 'If you can give me a purchase order number now I can get the goods to you tomorrow.' This is a trading attitude. It is something negotiators develop. It makes them more alert to possibilities. It ensures that they always get something in return for a concession. People who do not have a trading attitude often find negotiations emotionally uncomfortable, miss possibilities and concede too much.

'If'

'If' is a very small word but it is of tremendous value to a negotiator. It provides four real benefits:

- It links items together: 'I can … *if* you can …'; 'I could only agree to … *if* you could agree to …'.

- It makes everything you offer conditional on a trade.

- It enables you to test the customer's reaction to your ideas and proposals: '*If* I could … would you be able to …?'

- It subtly educates the other party that they will have to give something to get something.

'If' makes all your suggestions conditional on a trade. Remember, you might have something (tangible or intangible) that they value highly. They might have something you value highly that seems like peanuts to them. 'If' increases the chances that you will get it.

Tangibles and intangibles

Many people think too narrowly and assume that negotiation is only about tangibles such as price and quantity. These days, however, we deal as much with services as with manufactured goods and with knowledge and skills as much as products that fit in a box. Consequently, intangibles have become much more important. If a potential customer tells you that they were let down by someone in the past, and reliability is one of your strengths (remember the SWOT analysis?), your bargaining power has just gone up. You can give yourself much more room for manoeuvre if you broaden the topic under discussion. The 'something' you offer for trade might be *tangible*, such as the only widgets available with a left-hand thread or speed of delivery, or it might be *intangible* such as reliability, peace of mind, prestige and so on.

Most customers are as concerned with intangible issues – quality, reliability, honesty – as they are with the price.

If you combine your strengths (from the SWOT analysis) with the customer's wants, needs and motivators (discovered during your persuasive funnel probing) you are in a good position to link and trade creatively. Here are some examples of tangibles and intangibles to start you thinking.

Tangibles	*Intangibles*
• Cost	• Reliability
• Quality	• Confidence
• Variety	• Prospects of future business
• Timescale	• Kudos

Seed-sowing

When farmers and gardeners sow seeds they do so in the knowledge that there is a period when the seeds are growing underground and establishing roots before the plants gradually emerge and eventually bear fruit. Good negotiators do something similar. They sow seeds of thought in the other party's mind that will gradually germinate, take root and 'bear fruit' at a later stage in the negotiations.

Seed-sowing is an easy way of being a more effective negotiator.

There are two kinds of 'seeds' you might want to sow. The first kind is anything relating to a potential problem or perceived weakness. If your company is a relatively small player in the field, you might want to sow seeds of thought about how large suppliers can get bogged down in bureaucracy and lengthy decision-making processes, whereas small ones tend to be more flexible and swifter. You might want to sow seeds of thought suggesting that this customer will be one of many customers to a big supplier but a large and important customer to a small supplier and, hence, will get great service. The second kind of seed is anything relating to strength or your USP. If your company provides a longer guarantee than your competitors you might like to help the customer relate that to faith in the quality of your work compared to the competition. If one of your colleagues has recently spoken at an industry conference, you might like to raise that fact conversationally and link it in the customer's mind to your reputation and credibility.

There are two important points to note about seed-sowing. First, just like seeds that bear fruit long after they were planted, thoughts that you sow in a customer's mind seem to emerge from their own subconscious and are more readily accepted by them than points you make deliberately later on. Second, you will only know which seeds to sow if you have done a SWOT analysis and probed using the persuasive funnel. To prompt

Ease of supply from you

	Difficult	Easy
High	Play down Find an easier way of supplying it	Maximize Emphasize Trade well
Low	Ignore	Emphasize its importance Raise its perceived value Use as a concession

Value to the other party

Figure 3.1 *The negotiator's matrix*

your thinking, however, you can use the negotiator's matrix shown in Figure 3.1 to begin identifying what you have to trade. Remember, if you can identify something which the other party wants or values that you can supply relatively easily, your bargaining power has just increased!

Spotting and controlling manipulative negotiators

The ideas in this section will ensure that you become less susceptible to the tricks adopted by some negotiators. The main ones are described below so that you can recognize them and take action to counter them. However, it cannot be emphasized too strongly that you should never consider resorting to these tactics yourself as they do not lead to good working relationships.

Keeping you waiting

This is a tactic used by 'superior' people on 'junior' people. It is a way of saying, 'I'm the more important person'. They hope that you will either become more nervous (and consequently less effective) or that your schedule will become so pressing that you will agree to what they want in order to keep the discussion short.

Counter-tactics: Keep some work with you so that their attempt at pressure becomes a *gift of time* for you to use productively, or use the time for last-minute preparation. Alternatively, reschedule the meeting. If their delay was genuinely unavoidable, they will understand; if it was an attempt to manipulate you, they will see that it has not worked and be less inclined to try the same tactic on you in the future.

Your best offer

Picture the scene. You are a young sales rep and you've just entered Mr Big's office. You are no more than 30 seconds into your sales pitch when Mr Big's secretary comes through on the intercom with a pre-arranged message: 'Hello, Mr Big. Miss High-and-Mighty wants to see you in two minutes in her office.' What can Mr Big do, other than apologize and say, 'Well, let's skip the haggling. Just give me your best price'? What can you do other than go straight to your fallback price?

Counter-tactics: Whether you are negotiating a price for a product, the start date for a project, or how many staff you can temporarily second to another department, watch out if the other party puts you under unexpected time pressure and attempts to push you straight to your fallback position.

Try responding, 'I'd like to give you my best price but, until I've learned more about your requirements, I don't know what my best price is.' At best, that response will make Mr Big realize that pushing the sales rep is not the most sensible route to the best deal and, at the very least, it should let Mr Big know that he is not dealing with a push-over.

The principal ploy

Once you have learnt how to handle a negotiating ploy, it no longer works against you.

A favourite tactic in buying situations, the principal ploy is also found in multi-disciplinary project teams. Just as you are getting close to agreement, you discover that the person to whom you are talking does not have the authority to agree. They leave the room and return five minutes later saying that their boss will not agree unless another x per cent is conceded. They get the x per cent, disappear again, and return five minutes later saying that the boss now wants delivery (or whatever) in two weeks instead of four. The unseen principal always wants a bit more. Sometimes only the threat of consulting a principal is enough to stop the downward spiral.

Counter-tactics: Insist on discussing matters with the principal directly. Resurrect matters that the other party thought were already agreed. For example: 'If you want delivery in two weeks *and* an x per cent discount we'll have to take another look at quantity.'

The subsidiary decision

Here is an example. If you are unsure whether to buy a certain car or not, the salesperson may ask you which colour you would prefer if you were to buy one. 'Red? Let me see if we have one in stock ... Yes, I could have that ready for you tomorrow. Would you prefer delivery in the morning or the afternoon?' By asking you to make small decisions he is assuming that the big decision is already made. It is a technique favoured by pushy people. 'I need to borrow one of your staff. Would you prefer to lend me Bill or Sally?' assumes that the decision whether to lend a member of staff at all has already been made.

Counter-tactics: Remind them that the small decisions will only be made after the big decision. You are not refusing to help. It is just that you want to make the decision in your own way and in your own time. Use assertive language, good eye contact and a neutral tone of voice.

Last-minute wavering

Just when you think you have reached agreement, the other party begins wavering over a point because, as your relief at reaching agreement increases, your defences drop and they squeeze one last concession from you. At least you think it is the last one. In fact, they can waver several times, squeezing another concession from you each time.

Counter-tactics: Remember that every time they raise another issue, previously agreed points can be brought back for discussion using the word 'If': for example, 'I can consider this new point *but only if* we reconsider ...'. If the new point is genuine they will not mind resurrecting a previously agreed one; if the new point is not genuine, they will retract it.

An early concession

Some negotiators begin with an early concession and then wait for you to reciprocate and, being a lady or gentleman, you probably will – unless you realize that what they have done is the chess equivalent of sacrificing a pawn to take your queen.

Counter-tactics: Thank them, remember the concession for later and continue to explore.

Dragging a conversation on too long

When this happens the negotiator is hoping you will be tempted to agree to almost anything that offers the relief of a quick solution.

Counter-tactics: Allow plenty of time. Adjourn the discussion. Mentally 'lock on' to the outcomes you identified during planning, particularly your Batna.

Action points

1 Make a list of everything that detracts from your bargaining power and delete all the items that you could not easily justify to your manager. (If they can be deleted, it means that they exist more in your mind than in reality.) Discuss with your manager and/or your team how to tackle the others.

2 Make a list of everything that increases your bargaining power. Add more ideas from each chapter of this book. Plan how you can use them in a forthcoming negotiation.

3 Double the amount of preparation you normally do for a negotiation and assess the difference it makes.

4 Before a negotiation (or during a break) list your wants and needs and those of the other party. If you cannot easily list theirs, probe for greater detail.

5 Observe others in complex (professional) negotiating situations and see how they attempt to negotiate. See how many deliberately trade. Notice the effect on the customers.

6 When you are next preparing for a substantial negotiation, involve your team in a structured 'thinking hat' preparation session. At the end, review how useful it was and how they would make their use of the structure even more effective.

7 Watch an effective negotiator at work and count how many times the word 'if' is used and what effect it has.

8 At a team meeting preparing for negotiations, brainstorm as many tangibles and intangibles as possible. Review and prune the list and then decide how you can use it to enhance your bargaining power and trading, and to plan a deliberate 'seed-sowing' campaign before your next negotiation.

9 Ensure that all staff engaged in selling situations receive quality training in negotiating skills.

10 Take someone whose opinion you respect to a negotiation, and ask them to observe you and give you feedback.

Chapter **4**

Advanced Selling and Negotiating Skills

The previous two chapters have given essential guidance to help you win more business by being better at selling and negotiating. This chapter provides information that will further extend your selling and negotiating skills. The information offered is applicable to selling and negotiating face-to-face, and via the telephone or e-mail, as well as to written proposals and presentations.

First, it is necessary to review some essential points. When we are selling and negotiating we are effectively trying to gain a customer's agreement and commitment. When deciding whether to go along with us, they use *their* reasoning not *ours*. People's reasoning is a combination of rational thinking and emotional thinking.

Using these advanced skills in the right way, you can affect your customers' thinking – both rational and emotional. Approaching selling and negotiating this way dramatically improves your chances of gaining a customer's trust, being seen as a professional and, consequently, winning their business.

These advanced skills work so well because they are congruent with the way in which people's brains work. Research into neuroscience shows that the brain's limbic system, which controls our emotional thinking, is much more powerful than the neocortex, which controls our rational thinking. This means that our brains tend to give emotions priority over logic. So, in addition to appealing to someone's rational thinking, you can make your selling and negotiating even more effective if you also appeal to their emotional thinking. Described below is a range of ideas. Some appeal directly to emotions and some appeal to both reasoning and emotions.

Suggest, don't insist

This idea was mentioned as part of the persuasive funnel technique and is so important that it is worth mentioning it again here.

Make it easy for people to go along with you.

Interaction between two people exists in a series of triggers and responses. Behaviour from one person triggers a response from the other. Certain behaviours carry a high trigger–response predictability. For example, if you present your idea as a statement, it sounds like you are insisting, which triggers most people's *natural reaction* to resist. On the other hand, if you present the same idea as a suggestion, it becomes non-threatening and hence easier for the other person to consider. Here are some examples:

Statement	Suggestion
• *'What you ought to do is ...'*	• *'How would you feel if I suggested ...?'*
• *'What you have clearly failed to realize is ...'*	• *'Can I check what account you took of ...?'*
• *'You've obviously forgotten ...'*	• *'How does that relate to ...?'*
• *'I'm sorry, but I insist that we ...'*	• *'What would I have to say to convince you that ...?'*

As you can see, the statements are likely to trigger resistance whereas the suggestions, because they are more inviting, more involving and more thought-provoking, are more likely to trigger dialogue.

Embed your suggestions

Some suggestions, when spoken on their own, can sound like a presumption, an instruction or even a direct order! To make your suggestions more acceptable to other people you can *embed* them in phrases that make them easier to hear. Here are some examples. In each of them, I have underlined the suggestions and have italicized the phrases used to embed them.

If you want to say to a customer:

'Look at the figures in the table, especially those relating to maintenance costs, and it's obvious that ...'

but you feel that might sound unacceptably prescriptive, you could say:

'*As you* look at the figures in the table, *you might notice* that those relating to maintenance costs are ...'

Or you might want to say to a customer:

'I'd like you to <u>check our references</u>'

but, because you know that this doesn't sound very attractive, you could say:

'*If you want to, you can* <u>check our references</u>'

and this might sow the seed of thought in their mind that they do, in fact, want to do this.

In Table 4.1 you will see a range of embedding phrases together with an explanation of the rationale behind them.

Embedding phrases	Rationale
'You might already know that ...'	'There is no need for you to disagree with this point because I'm not forcing it on you. All I'm doing is reminding you of something you already know.'
'Eventually ...'	Eventually, everything happens. Eventually means inevitable. So if eventually everyone comes to realize (whatever you are saying), as long as your tone of voice remains respectful, your audience is more inclined to accept it.
'I could tell you that ... but you probably already know.'	Effectively, you are saying, 'I could tell you [this point] but I'm not going to. So there's nothing for you to resist or to disagree with. Consequently, your audience is more inclined to relax and listen.
'You can ...' 'If you want to you can ...'	'I'm not asking you to do something, I'm just letting you know that its OK for you to do it.'
'I'm wondering if you'll ...'	Again, you're not actually asking for anything, you're only thinking aloud, so there is nothing for your audience to resist as they might do if you demanded or even asked directly.
'Can you imagine the [benefits] ...' 'You might find that ...'	Again, you're not asking anyone to agree, disagree or to commit themselves – that can make people feel uncomfortable. You're simply asking them to think about something. And if, in doing so, they realize the benefits of what you're suggesting, they're more likely to go along with it.

Table 4.1   *Embedding phrases and their underlying rationales*

Here is a longer example. Again, the suggestions have been underlined and the phrases used to embed them have been italicized. You might find it easier to see the difference if you imagine a change in tone of voice between the suggestions and the phrases used to embed them. The (underlined) suggestions can be spoken deliberately and clearly and the (italicized) embedding phrases can be spoken more gently.

> Here is an idea. *As you* mentally rehearse something you want to say to someone, *you might like to* try 'embedding' some of the suggestions you want to make. *You can* think of the suggestions you want to make to the other person and then choose some appropriate embedding phrases. *You might find that* what you say feels different, that it has a gentler, more inviting tone to it. *To help* you feel comfortable with embedded suggestions, *you can* practise in social situations before you try them out in more formal settings such as business presentations or meetings. *Eventually*, you will feel more comfortable with these phrases, and they will come more naturally to you. *If you want to, you can* go back over this paragraph and identify the suggestions I have embedded in it for you.

Embedding suggestions works in face-to-face situations, telephone conversations, e-mail communications and written correspondence, written proposals and presentations.

Use their words, phrases and metaphors

Here is an interesting fact. While most people have a 'recognition' vocabulary of 20 000 words, they only use a 'day-to-day' vocabulary of 2000 words. So, from the vast range of words available to us, we choose to use about 10 per cent regularly. We choose the ones with which we are comfortable. So as you listen to customers (which you will be doing naturally at the wide part of the persuasive funnel) consciously spot the words, phrases and metaphors they use and then deliberately use the same words, phrases and metaphors back to them, especially those relating to key points or priorities. In doing so you will appeal to both their rational thinking and their emotional thinking.

Use the same language as your customers and you will gain their trust.

This way the customer is subconsciously reassured that you are listening to them (which means they are more likely to listen to you) and what you say will automatically sound more comfortable to them.

Here is an example:

You:　　　So, if we can have your staff for half a day, we can train them to use the sophisticated security equipment after the building is complete.

Customer:　I don't see that happening. We've got some tough targets to achieve. I can't have them gallivanting off on some training course. I've got to show results, you know. With me, it's invest not spend.

You:　　　So, if it's important that your staff achieve *tough targets* but you can't *do without them* while they're on *some training course*, what results would you need to see to feel confident that the course was a worthwhile investment?

Here is another example:

You:　　　We know from the last work we carried out on this site that local residents object to the noise and may complain to your staff.

Customer:　That's your job. My troops are in the firing line enough these days. I can't have them confronting members of the public. They're completely unprotected.

You:　　　If you're concerned about your *troops* confronting the public, let's see what we can do to *protect* them when they're in the *firing line*.

Use positive and negative words to 'steer' customers' thinking

Have you noticed how some words and phrases trigger a negative feeling while others trigger a positive one? For example, a *swift response* is preferable to a *knee-jerk reaction*; points about *value* are easier to hear than points about *cost*. Management talks about *modernization* and *restructuring* whereas unions and others refer to *cuts*. You can use different words to describe the same thing depending on the direction in which you wish to steer someone's thinking.

So, as you sell to and negotiate with customers, you can use negative words and phrases to help 'steer' the other person *away from* what you don't want them to do and you can use positive words and phrases to 'steer' them *towards* what you do want them to do.

Here are some examples of words identified as being more subconsciously influential than others (where appropriate, with some 'negatives' next to them). As you read them you will probably recognize many of them from advertisements, commercials and politicians' speeches. They use them because they work.

- you/yours
- positive (negative)
- greatest
- new (old)
- tested (untested)
- invest (spend)

- easy (difficult)
- discover
- proven (unproven)
- guaranteed (risky)
- unique (common)
- certain (unpredictable)

- free (costly)
- success (failure)
- results
- safe (risky)
- better (worse)
- bonus

Build natural momentum into the conversation

In addition to using words and phrases that have a positive feel, you can stimulate a *natural momentum* in the other person's thinking so that your suggestions feel natural and reasonable. For example, you can:

- Focus on their motivation. *'How keen are you to ...?'* implies they are already keen, it's just a question of how keen they are.

- Help them focus on tangible results. *'What benefits would you like to see if we do implement this idea?'*

- Encourage them to modify your suggestion, rather than come up with an alternative proposal. 'If that idea doesn't quite meet your needs, *how would you fine-tune it?'*

Summary

We have looked at a range of ideas that will help you win more business by advancing your selling and negotiating skills. There are, however, two words of caution.

First, information can enter someone's brain consciously or subconsciously. When it enters consciously, you know it and you can respond accordingly. When it enters subconsciously, you do not know it is there, but you *experience the feeling* that it has triggered. The techniques described above only work when they enter the other person's brain subconsciously – which means that you have to use them subtly and gently. If the other person notices you using them, not only do the techniques lose their effectiveness, but they could also provoke a negative reaction in the other person. Indeed, if you yourself have not warmed to these advanced techniques, this might explain it; you know what this chapter is about, and the techniques have been explained to you in detail. This means that they have entered your brain consciously. However, imagine if the techniques were used sparingly in the context of a business conversation, you would be unaware of them at a conscious level but would experience the feelings they trigger at a subconscious level.

Don't try to manipulate customers. It always backfires in the end and it could cost you dear.

Second, all these techniques can be used manipulatively. I firmly believe that manipulative influencing backfires sooner or later and disadvantages the manipulator. Your only defence, therefore, is to always treat people openly, honestly and respectfully.

Action points

1   Reflect on how you talk to staff, colleagues, managers, friends and family, especially when you are giving advice. Do you use autocratic language (*'What you ought to do is ...'*) or suggesting language (*'What would be the effect if you tried ...?'*) Practise using more of the suggesting language and observe the effect.

2   Try saying something without embedding and then embedding your suggestions and see what the different effects are.

3 Deliberately listen to the words and phrases people use and, very subtly, use some of them in your response. See what the effect is.

4 When you are next trying to convince someone of a course of action, deliberately use positive and negative words to steer them in the direction you want. Again, see what the effect is.

5 Do the same with momentum-building phrases.

6 Choose one skill from this chapter and deliberately use it several times a day for a week and see how natural it begins to feel.

Chapter **5**

Sales Proposals and Presentations

In this chapter we are going to look at the importance of proposals and presentations to winning new business. We'll look at some of the problems and then at a range of proven ideas that will help differentiate your proposals and presentations from your competitors.

Background

Written proposals and formal presentations are often important, business-winning steps in the construction industry tendering process. They are so intertwined that they can be covered most effectively in one chapter.

This part of the book will show you the critical elements of proposals and presentations and the techniques you can use to gain advantage. It will also include the importance of the traditional SWOT analysis and, to help you further, introduce a new model relating your company's strengths and weaknesses to a customer's needs.

The growth of outsourcing, benchmarking comparisons and compulsory competitive tendering means that many construction companies have to pitch for business. It has also led to a growth in formal tendering procedures, often via a professional purchasing department. Pitching for business in this context has several components:

- a **supplier** who is making the pitch.

- the **pitch** – a proposition or offer that the supplier hopes the customer will find attractive enough to award the business to them. The pitch might be in response to an invitation to tender or pre-emptive to secure a contract renewal or entirely speculative.

- a **customer** who will consider the pitch, often by comparing it to comparable pitches from your competitors. Sometimes, the customer has two sections you need to address – the people who initiated the request and the purchasing department who compiled the invitation to tender. Purchasing departments sometimes review tenders rationally (they may not even be construction experts), but the initiators are likely to review tenders both rationally and emotionally. The information in this chapter is designed to get you through the purchasing department's review so that you can then impress the initiators.

Let's begin by looking at the three main problems when pitching for business.

Pitching problems

'Pitches' can be close to, or wide of, the mark. Either way, a poor pitch is one that fails to clarify how you will meet the customer's needs and selection criteria. A quick 'topping and tailing' of an existing proposal by a rushed executive often falls into this category.

Another poor pitch is one that confirms that you can do the job but fails to distinguish you from the competition. You are unlikely to be the only company the customer has invited to tender and, unless they are guilty of the point above, they will all meet the customer's criteria.

Finally, if your written proposal gets you shortlisted, you are invited for interview or to make a presentation, and all you do is make a standard 'credentials presentation' – or duplicate the written proposal – perhaps with a few visual aids to brighten it up.

The ideal 'pitch' is one that:

- confirms you can meet the customer's criteria as a minimum
- proves that you will be great value to them
- distinguishes you from the competition
- excites or motivates them in some way by highlighting benefits they had not thought of themselves
- wins you the business.

That means you need to prepare your pitch broadly, rather than narrowly.

Preparing broadly

In 'pitching' situations it is easy to think and plan too narrowly, thereby omitting easy-to-supply features of your offering that

Customers will only know how they benefit if you tell them.

the customer would find attractive (refer back to Figure 3.1, 'The Negotiator's Matrix'). Thinking too narrowly is usually a result of designing a proposal to meet the customer's specification and then simply highlighting the first-level benefits to them (see below). An alternative approach is described below. As you consider it, bear in mind that preparing a 'pitch' is an investment and, like all investments, what you expect to gain needs to be worth the time, effort and money you put into it. Proposals for small contracts, therefore, are bound to be more modest than proposals for large contracts. The basic steps are as follows:

1 Seek information from a wide range of sources:

   - people with knowledge of, or a relationship with the customer
   - people with experience of the important technical aspects of the proposal
   - people with experience of the commercial aspects of the proposal.

2 Identify the customer's essential and desirable criteria. The obvious criteria will be in the customer's invitation to tender. Because customers are familiar with their environment, however, they have may have taken some essential criteria for granted. So also seek information from anyone who understands the customer and/or important issues in their environment.

3 As you consider the benefits that the customer can gain from what you have to offer, think one step beyond normal. For example, imagine that you would normally think about how one of your strengths/features/USPs would provide a benefit for the customer as follows:

Feature	*Benefit*
Our buildings have won architectural awards which means that ...	... you will get a prestigious new head office.

Now, go beyond that to the next level and think what that first benefit would do for the customer:

Feature	*1st level benefit*	*2nd level benefit*
Our buildings have won architectural awards, which means that …	… you will get a prestigious new head office, which means that …	… city investors will see that you are serious about your recently announced plans for growth.

This will set you apart from your competitors who will stop at the first level of benefits. You can only take this step, however, if you have used the persuasive funnel, considered the customer's needs, as well as their wants, and sought information from the wide range of sources mentioned in step 1 above.

4 Identify anything and everything that you, as a supplier, are good at, that has impressed other customers and of which you are particularly proud. Make sure that you include intangible items. (You can, of course, do this much more easily if your SWOT analysis is up-to-date.)

5 Consider all the information collected so far and brainstorm an initial list of anything and everything that a proposal would need to include to be successful.

6 Next, arrange the list in matrix form, as illustrated by Figure 5.1.

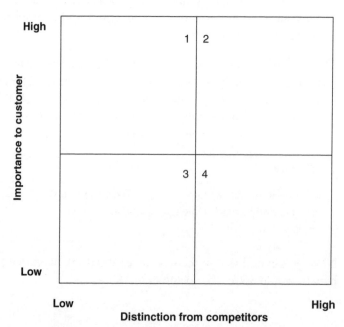

Figure 5.1 *The proposal planning matrix*

7 Examine the matrix and plan how to:

- promote quadrant 1 items to quadrant 2 by describing the value the customer will receive as a result of the *way* you provide those items. As the item itself is not a USP, emphasizing the way you provide it might be a vital distinguishing factor.

- promote quadrant 4 items to quadrant 2 by emphasizing the benefits to the customer of those items. If you can't find any first-level benefits go to the second level.

You may always politely decline the offer to pitch or quote for business. But make sure that you leave the door open for the client to come back to you with other work in the future.

However, be objective and realistic. If you genuinely cannot meet the customer's essential criteria, you have only two alternative courses of action. First, check that the customer's essential criteria are genuinely essential and not just highly desirable. Unlike essential criteria, desirable criteria can be negotiated or compromised, yet it is surprising how often people confuse the two. Second, consider whether you really want to invest time, money and effort in what might prove to be an abortive pitch.

8 Similarly, double-check that you can, in fact, deliver everything you might be promising. Otherwise, the contract may become a burden and cost you dearly in penalties.

9 Prepare your pitch (written proposal or presentation) in such a way that real and 'tailored' benefits to the individual customer are clearly presented. Also, by considering the audience characteristics very carefully you can pre-empt potential objections. You can draw attention to the issue and, as part of the presentation, explain how your recommendations, strengths and so on address it. For example, if price is a potential objection for the decision-makers, explain how quality may cost more initially but lack of quality will cost much more in the long run. If the decision-makers feel that doing something new is risky, explain how being first off the blocks enables an athlete to gain the inside track and so on.

10 Get some colleagues to play devil's advocate and find all the faults and weaknesses in your pitch. Then, and only then, are you ready to prepare a written proposal or presentation.

Adopting this approach has several benefits:

- It saves you investing time, money and effort in an abortive pitch.

- It clarifies what is relevant and therefore should be included in the pitch, as well as what is not relevant and therefore should not be included.

- It ensures that a pitch focuses on an individual customer's needs rather than on what you might want to sell or might have found successful in the past.

- It improves your chances of a successful pitch by distinguishing you from the competition.

- It leads to greater customer satisfaction because you can ensure that your staff supply what has been promised.

Once you have collated all this information you are ready to prepare a written proposal or presentation.

Written proposals

Here are some of the common faults with written proposals. As you read them, you can quickly work out what a good proposal should look like and the information it should contain.

- Rigidly sticking to (at best) your own house style or (at worst) the proposal format you have always used.

- Topping and tailing or cutting and pasting previous proposals but doing so in a way that 'the joins show'.

- Failing to provide the information requested in the invitation to tender or, if there is no written invitation, the information the customer has otherwise requested.

- Providing only the information the customer has requested.

- Providing poor presentation in terms of factual errors, poor spelling and grammar, typographical errors and lack of visual appeal.

- Sending it too early or too late.

- Not following it up with a telephone call.

When you re-use material from an earlier proposal, make sure that you take time to adapt it for the new client.

So, bearing these points in mind, here are some guidelines for written proposals.

Style and structure

Having a house style – hopefully a good one – is one thing but if the customer has listed their requirements in a certain sequence, other companies will probably follow it. This means that you, too, need to stick to the sequence to make it easy for your customer to compare your proposal to others. Incidentally, it is a good idea to summarize the customer's need, your company's background and why they asked you to tender. Also, although you may need to include some statistical information in the body of the proposal, putting too much in will spoil the document's flow so you might want to consider locating large amounts of statistics in an attachment or appendix.

Cutting and pasting

Whilst there is no point in reinventing wheels, proposals need to look fresh and tailored. So, although you might legitimately copy aspects across from other proposals, it is a good idea to get someone to go through it with a fresh pair of eyes to ensure, amongst other things, that it flows. (This is another of the benefits of step 10 above, because authors rarely see their own mistakes.)

Content

Written invitations to tender normally specify the information that has to be provided. If you rely on your own company's house style for proposals, you might miss some small, but important, point of information. It is a good idea, therefore, to use the invitation as a checklist. If there is no written invitation to tender, hopefully someone will have followed the persuasive funnel to uncover the customer's criteria, wants and needs and will be able to use that as a checklist to ensure that the proposal covers all essential information.

Incisive content

A written proposal has to answer questions even though the customer or its purchasing department may not have asked them specifically. All they may ask for is information – so you need to ask why they want that information. For example:

They may want ...	... which means they're asking
An organization structure	Have you got the internal resources to handle a job this size?
Several years of financial accounts	Will you be around long enough to finish this job? Will this job stretch you too far financially? Are we going to be a large, important customer for you or a small, insignificant one?
Examples of similar jobs you have worked on	Can we be confident in you?
CVs of key people who will work on the job	Can we be confident that the project team won't be headed by the graduate trainee?

By providing the information *and* answering the implied questions, you are distinguishing yourself from the competition.

Visual presentation

Another way of distinguishing yourself from the competition is with the proposal's presentation. Hopefully, it goes without saying that it should be free from factual, grammatical and spelling errors, but what about its visual appeal? Here are some ideas to help you.

Take time over the layout of your proposal. it is easy to lose a contract not on the substance of your offer, but because it contains spelling mistakes.

1 Leave good-sized margins at the sides and top and bottom of the page. Feel comfortable about leaving white space. Pages should never look crowded.

2 Use headings and subheadings to help the reader find their way through the information and to effortlessly know what is a main point and what is a subpoint.

3 Reinforce this structure with both a numbering system – for example, 1.0, 2.0, 2.1, 2.2, 2.2.1, and so on – and indentation so that subpoints have a wider left-hand margin than main points.

4 Explain in the introduction how the proposal is structured. This sets up the 'pigeon holes' in the reader's mind so that none of their attention need be diverted to organizing the information. That way, you get more of their attention.

5 Finish with a summary that emphasizes key features and benefits.

6 Ensure that the author's name and contact details are prominent, in case of queries.

7 Finally, if the customer wants hard, as opposed to electronic, copy, make sure that it is attractively and securely bound. If the words in your proposal are saying that you are experienced, professional, secure, reassuring, not cheap but excellent value and so on, your binding has to say the same thing.

If you use these ideas you will find that your proposals have more business-winning potential. You might, however, also have to present them and, as this is a major opportunity to differentiate your company from the competition, the more powerful your presentation the better.

Powerful presentations

What is a powerful presentation? One in which you dress in white and have them rolling in the aisles? No. It is one in which you impress the audience with your professionalism, thereby distinguishing you and your company from the other presenters and their companies.

Potential customers subconsciously equate your presentation ability with your professional ability.

It is a well-proven fact that how we visually appear to, and interact with, decision-makers affects their assumptions and hence the decisions they make about us. This is not to suggest that a slick presentation will compensate for a lack of technical expertise. (Even if it did, the deficiencies would soon become evident.) It is to suggest, however, that, in the decision-makers' minds, your technical expertise is probably as good as that of everyone else who is pitching for this business.

When you add to this the fact that many people dread public speaking (the majority of people, apparently, are more afraid of public speaking than they are of dying), a little knowledge of what makes a powerful presentation will not only remove much of the fear, it will differentiate you from your competitors. The information that follows will enable you to structure a presentation and give it some 'oomph' – and to do so more easily than you thought possible. Here you will learn the characteristics of powerful presentations, the rationale behind them and then some tips to help you further.

Mind-set

Powerful presenters are positive. They expect to do well and to enjoy doing it. Even if they are not really looking forward to the presentation, they 'pretend' they are. Nerves and anxiety often show in our behaviour. During presentations the behaviours you notice are uncomfortable body language, hesitant speech and apologetic terminology – for example, 'This is just a little slide to show you ...' rather than 'This next point is important, as you look at this slide I want you to notice how ...'. The audience wants to enjoy the presentation and are much more likely to do so if you also seem to be enjoying it.

Structure

Powerful presentations have a discernible structure. There will be a main, recurrent *theme* in the form of a *single compelling message*. There will be *main points* from which spring *subpoints. Exceptions* and other additional information will be clearly distinguished. One point will lead naturally to the next so that the structure 'flows' in much the same way as a cascading waterfall. When human beings receive information their brain automatically attempts to structure it. So, if you don't structure the material and explain the structure to your audience, their attention will be reduced as part of each listener's concentration is diverted to the structuring process.

If, on the other hand, the presenter explains the structure clearly (for example, 'OK, now to my second point which, as I explained, is slightly different from the first') the message is received more consistently and attention is greater.

Presentation techniques

Powerful presenters use a few simple, but effective, techniques or 'tricks of the trade' that have stood the tests of time and of common sense.

- Their opening is *thought-provoking* and attention-grabbing. All audience members have an unasked question: 'Why should I listen to this?' Opening with a thought-provoking statistic, question, observation or reference to the audience's goals answers that question and engages their attention.

- Their opening is *physically* engaging. Powerful presenters speak to each member of the audience. They do so by appearing relaxed, smiling and, in particular, by making eye contact with every person there.

- They provide overview so that the audience understands the structure (see above).

- They deliberately link one section to the next so that, although it is a separate point, the presentation still flows. Links not only assist structure, they also help the audience assimilate the information effortlessly.

- They encourage active and/or passive participation. Participation retains an audience's attention and appears to 'personalize' the presentation for each audience member. *Active participation* means that you encourage the audience to join in by asking them questions (see 'Presentation style' below) or, if you are demonstrating something, getting them to join in. *Passive participation* means that you encourage them to join in *mentally* by asking them rhetorical questions, by using phrases that 'hit the nail on the head' (in other words, relate to the senses) and by making suggestions that deliberately trigger thoughts (for example, 'So, think of the last time you were ...' Or 'Can you imagine what the effect will be when you ...?')

Presentation style

A powerful presenter's style is usually relaxed and businesslike but, above all, it is genuine and personal. That means:

- They make good eye contact with everyone, including those people seated on the extremes. Eye contact tells people that you are talking to them personally, so you need to look at each of them.

- Frequently, they enter the audience's territory rather than staying stuck behind their lectern or top table. By moving into the audience's territory, you establish more of a bond with them. The alternative is to appear 'stand-offish', distant or in need of security.

- Their style is 'inviting', 'empowering' and informative. The audience thinks, 'I never knew that.' Whether asking questions directly or rhetorically they rarely put the audience on the spot. The audience needs to feel 'safe', that it's OK to join in, that you are a nice person. They can only do that if you are helpful and positive.

- They speak 'from the heart' rather than from a prepared speech. They often speak (apparently) without notes. This is important because, as we tend to speak much more informally than we write, reading from a written document usually sounds formal and 'stiff'. That applies both to a prepared speech and to detailed notes. Although all presenters should know their material inside out, they also need prompts, so why not use your visual aids as both an aid for your audience and as a prompt for yourself?

- Their pace is swift. Some presenters slow down when delivering a presentation in the mistaken belief that it will be easier for the audience to understand. That is rarely the case because, when listening to a slow talker, we find it more difficult to discern the structure and consequently our attention wanders. Powerful presenters, therefore, deliver their material at a fast pace but, because of the clear theme, structure and links, their audience keeps up with them easily.

There's a fine line between a fast and clear presentation and a garbled one that leaves customers confused.

Visual aids

Powerful presenters follow the KISS rule (Keep It Short and Simple) and use simple, yet effective, visual aids. Complex visual aids are confusing and they distract people's attention from you and your message.

- Their visual aids carry no more than six points. Using more than six points means using smaller lettering to accommodate them on the slide and they become too crowded, too complex and too difficult to read.

- They prune all non-essential words from their visual aids. They add the other words with their voice. That way, the visuals stay clear.

- Their visual aids contain simple icons, sketches, matchstick people, symbols and so on. A picture may speak a thousand words, but it still needs to be a simple picture.

- Their visual aids are arranged so that people's eyes follow them naturally. In Western societies, because of the way we read, our eyes tend to track a 'Z' from the top left-hand corner to the bottom right-hand corner of a visual aid, so the top left and bottom right corners are great positions for key words.

Preparation

Powerful presenters prepare thoroughly. That is why what they do looks so easy.

- They practise a lot in their imaginations. Not only are mental dress rehearsals as effective as real dress rehearsals, you can do them anywhere.

- They work from the simple to the complex. They start with the theme, then the main points, then the subpoints, then the exceptions and then the links. This means that they have to remember less because one point leads automatically to the next. With structure and links, you can remember much more and recall it more easily.

- In addition to checking the venue and all the equipment, they will check the view from where the audience will sit, especially those who will be seated at the extremes. They want nothing to distract from what they say.

- They focus their presentation on the audience's crucial point, WIIFM – 'What's In It For Me?' Whether the purpose of the presentation is sales, information dissemination or education, a key factor in audience perception is 'Not only was there something in it for me, it was more than I expected'.

Tips on public speaking

In addition to the characteristics listed above, people who appear adept at public speaking employ three other techniques.

1 **Sensory specific language.** Information reaches our brains via our senses. Information can also trigger emotions. Powerful speakers capitalize on these facts. They liven up their message and make it register much more forcefully by using sensory-specific language. Their words will cause you to imagine what it will look like, see what they mean and understand what it feels like. What they say will strike a chord with you, ring a bell or even remind you of the 'sweet smell of success'. Their message will reach the four

corners of the world, echo in your mind and receive loud and resounding applause. This is not the kind of language you would find in a report or proposal, which is probably why it makes such a difference when public speaking.

2 **Threes.** We tend to warm to things that come in threes such as 'the Father, the Son and the Holy Ghost' or 'The good, the bad and the ugly'. Public speakers find using this device helps make their case more acceptable to their audience.

3 **Alliteration.** Similarly, a sequence of words beginning with the same sound also seems to register with us. Which all goes to prove that what you say does not have to be big, bold and beautiful for it to be as attractive as a warm welcome on a winter's night!

Note. Here is an important word of caution, however. When preparing a meal, a cook knows that some spice and herbs will make the difference between an average meal and a great one. They also know that too many herbs or too much spice will turn an average meal into a dreadful one. It is the same with these three tips. Overuse them and your presentation will be pitiful; use them sparingly and your presentation will be powerful.

Summary

The growth in outsourcing and in the use of information technology means that sales presentations and presentations to support proposals are probably more frequently used than they used to be – and many of them are being made by technical specialists rather than by trained salespeople. Using the ideas explained in this section, you can make your presentations more powerful, more impressive and more successful.

Success often depends on the presenter's ability to relate their proposal to the benefits the customer will receive. The relationship between the presenter's recommendations and the customer's needs has to be clear and simple – *from the decision makers' perspective*, not just from that of the presenter.

That clear and simple relationship is achieved most easily when the presenter summarizes the customer's situation, then the issues or problems and then explains how their recommendations will resolve the problem.

By considering the audience characteristics very carefully (in addition to any pre-presentation research that can be done) you can also pre-empt potential objections by drawing attention to the potential problem (objection) and, as part of the presentation, explaining how your recommendations address that issue. For example, if price is a potential objection for the decision-makers, you might explain how quality may cost more initially but how lack of quality will cost much more eventually. If doing something new might feel risky to the decision makers, you will explain how being first off the blocks enables an athlete to gain the inside track and so on.

Action points

1 Give a new manager a week, as part of their induction, to review a selection of written proposals against the advice in this chapter, to recommend improvements in your company's proposal writing and to run a training course on the subject. Make attendance compulsory for all proposal writers.

2 Use the proposal planning matrix (Figure 5.1) as an exercise during your next team development event.

3 Ensure that everyone involved in presentations to support a proposal receives good-quality presentation skills training.

4 Set up a reciprocal agreement with another department that you will play devil's advocate on each other's proposals.

5 Invite managers from other departments to attend your sales presentations, review them against the advice in this chapter and recommend improvements.

Chapter **6**

Winning New Business Through Customer Care

In any book about winning new business the obvious and exciting ideas to include are those that will help you improve your marketing, selling, negotiating and proposal skills. Yet, very often, the ways in which we can win new business – and sometimes lose it alarmingly easily – are neither quite so exciting nor quite so obvious. But they are real and important, nonetheless.

So, in this chapter we are going to look at how you can retain and win business through customer care. In it, we will cover some fundamental facts about customer care, address some of the inaccurate perceptions you might find in your company about customer care and explain what you can do about them.

Customer care fundamentals

Let's begin by looking at some customer care fundamentals.

What is customer care?

Customer care is a way of interacting with customers. It is based on a genuine appreciation of their custom that shows in company policies and individual staff members' behaviour. These combine in a way that meets customers' basic and human needs during the transaction.

Customer care is the simplest and cheapest way of differentiating your company from your competitors.

If that sounds a bit formal, you could think of customer care as a way of interacting with customers that leaves them so satisfied at the level of service that they not only want to do business with your company again, they also want to recommend you to others.

If that *still* sounds a bit formal, how about '*it's one of the simplest and cheapest ways of differentiating your company from your competitors and winning new business*'? First, however, you have to understand your customers' needs.

Basic and human needs

When customers interact with a supplier their needs can be divided into basic needs and human needs. Basic needs relate to the essence of the transaction such as buying the ticket, having the computer installed or having the illness diagnosed. Human needs relate to how they feel as a result of the transaction – for example, whether they feel appreciated, valued, reassured and so on.

It's surprising how important human needs are.

Customers' basic needs will vary depending on the transaction taking place. A customer whose basic needs relate to a kitchen extension will vary from a customer whose basic needs relate to the construction of a bypass. Their human needs, however, can be remarkably similar. The key point to note is that building a good kitchen extension or an excellent bypass does not in itself guarantee customer satisfaction. You have to meet their human needs as well.

A good example is that of visiting the dentist. Your teeth could have been expertly checked, filled and cleaned but, despite the discomfort, the dentist could still leave you feeling happy with the service as a result of the way in which he or she has treated you. Looking at another example, you could buy a new car and go to the garage to collect it (an exciting occasion for most people) but the salesperson could still leave you feeling dissatisfied because of the way they treat you. Fulfilling one set of needs is not enough to guarantee customer satisfaction; you need to fulfil both basic and human needs to get a satisfied customer.

So what are the basic needs and human needs? Here are some ideas to start you thinking. Once you understand them, you can begin to consider to what extent they are satisfied by your company.

Basic needs

Basic needs relate to the basic expectations a customer has when making a transaction. They include:

- a product or service that does what is expected of it
- efficiency in the provision of the product or service
- honesty and fair treatment
- courtesy
- full attention at all times
- a clean, tidy and welcoming environment

- an environment that looks efficient and businesslike

- not to be kept waiting

- being kept fully informed of what is happening

- kept promises

- someone who doesn't take complaints personally.

Looking at these basic needs, you can probably see how failing in any of them would dissatisfy customers. You can probably also see that getting them right is unlikely to be sufficient to satisfy them. That is where the human needs come in.

Human needs

Human needs can be subdivided into emotional needs, reassurance needs and esteem needs. They include such things as:

- personal warmth

- to feel good about themselves as a result of contact with you

- to gain a sense of achievement

- to believe that you enjoy dealing with them

- having a problem replaced with a solution

- to feel confident in you

- to feel secure and free from anxiety relating to the work you are doing for them

- to believe you understand their needs and are on their side

- to feel they and their custom matter to you

- to be treated as important.

Understanding the many and varied needs of customers enables you to identify the needs most appropriate to them and to take action to satisfy those needs. (A selection of actions appears at the end of this chapter.) But a perfectly legitimate question to ask in a book about winning new business is, 'Why bother?'. The answer is: because it is a major factor in retaining customers.

Retaining customers

It is a well-established fact in most business sectors that retaining a customer is significantly more cost-effective (profitable) than replacing a lost one. Add to that the old adage, *'once a customer always a prospect'*, and you can begin to see the significance of customer care. Let's start by looking at the cost of losing a customer or, as they say in some circles, the cost of a high customer churn rate.

Retaining customers adds to the bottom line.

We will take two examples. We will calculate the cost of maintaining 100 customers, first, where the churn rate is 10 per cent and, second, where the churn rate is 50 per cent. We will assume, modestly, that it costs five times more to attract a customer rather than retain one.

No. of customers	'Churn' rate (%)	No. of customers needed to get back to 100 (% of column 1)	Cost of attracting new customers (to get back to 100) = £5n per customer (Column 3 x £5n)	Cost of retaining existing customers at £n per customer (Column 1 - column 3 x £n)	Total cost of 100 customers (Column 4 + column 5)
100	10	10	10 x £5n = £50n	90 x £n = £90n	£140n
100	50	50	50 x £5n = £250n	50 x £n = £50n	£300n

So, as you can see, if a business with 100 customers does not care for them and loses 50 per cent of them, it costs substantially more to maintain the customer base than the company that practises customer care. Suppose, however, that this average cost of it being five times more expensive to replace a customer than to retain an existing one does not apply to the construction industry. Suppose it is nearer 15 times more expensive. What happens to the cost of customer churn then? Let's apply this new figure to the same example.

No. of customers	'Churn' rate(%)	No. of customers needed to get back to 100 (% of column 1)	Cost of attracting new customers (to get back to 100) = £15n per customer (Column 3 x £15n)	Cost of retaining existing customers at £n per customer (Column 1 - column 3 x £n)	Total cost of 100 customers (Column 4 + column 5)
100	10	10	10 x £15n = £150n	90 x £n = £90n	£240n
100	50	50	50 x £15n = £750n	50 x £n = £50n	£800n

As you can see, even though the churn rates are the same, the higher costs of attracting new customers has a dramatic and costly effect on the business.

Remember, these figures are based on 100 customers. If your company has 1000 customers or 10 000 customers, these figures rise alarmingly. And, remember, this money comes straight off the bottom line! Working hard to win new customers in that situation is a bit like trying to make a yacht with a huge hole below its water line go faster by hoisting more sails. Hence its importance in this book.

There is another side to the coin, however. If your customer churn rate is low, you gain a double benefit – fewer customers to replace *and* greater lifetime value from each retained customer.

Lifetime value

Lifetime value is the worth of a customer to your business throughout the time they remain a customer.

By way of example, have you ever calculated your lifetime value to the garage that services your car or the supermarket where you do your weekly shopping? If you spend £500 a year on car servicing and keep your car for five years you are worth £2500 to that garage. If you stay loyal to that marque and that garage for, say, 25 years, you are worth £12 500 to them. Estimate how much you spend on average each week in your local supermarket. Multiply that figure by 52. Multiply that figure by 40 years and you will see that you could be worth well over £200 000 to that supermarket. That is a lot of money for them to lose if someone upsets you while you are trying to buy a loaf of bread one day.

How many of your customers have come back to you for more construction work over the years? Isn't that business cheaper to win?

Have you ever calculated how much your customers spend with you and how much they would be worth to your company if they remained customers for many years? As long as you continue to give good service (both basic needs and human needs) they will remain customers and, even if you have to compete with other firms as contracts come up for tender, you will have the inside track. Customers need very strong reasons to move away from suppliers who look after them, but are always eager to move from one who does not provide good customer care.

Without doubt, the business case for customer care is a strong one but what are some of the trends you need to be aware of?

Trends in customer care

Listed below are the main trends in customer care. They will not all apply directly to construction, but it is important that you are aware of their effects because they are changing the standards across the board. These trends raise the bar for everyone because people tend to apply the customer care standards they experience in one situation to other, different situations.

1 'Service' is overtaking price as the dominant criterion for many people, with 55 per cent of customers regarding service as more important than price. Whilst this may not always apply to the purchasing departments you have to deal with, as a social trend it points to the growing significance of the intangibles we discussed in Chapter 3 on negotiating.

2 Customer expectations are rising due to the *ripple effect*. In other words, improved customer service standards in one industry, such as banking, raise our service expectations from other sectors, such as construction. Consequently, nearly 90 per cent of UK consumers demand better service than they did five years ago and over 60 per cent will change suppliers if they don't get it. Whilst you may not be dealing with consumers in a conventionally accepted sense, the people in your customers' offices are consumers when they deal with their local supermarket, garage or utility company. They won't lower their standards just because they are now dealing with you. The standard ratchets up.

3 Customers have less patience than they used to. We use IT to get answers faster than we used to. We use e-mail to communicate with people in seconds instead of days. We use telephone or web-based services (for example, for banking transactions) to do more of what we want when we want to do it. We use mobile phones to access people more quickly than we used to. Having now become accustomed to rapid responses, our expectation of instant response has become so high and our tolerance of delay so low that *speed has become a competitive tool*. The speed with which you and your colleagues respond to e-mails, voice-mails and so on has suddenly magnified in importance.

4 We have become more of a 24-hour society than we used to be. Not only can individual customers, for example,

transfer money from one account to another or buy groceries in the middle of the night, companies will process information in another time zone to have it ready by the start of the following working day. More people are working in global companies with global customers. Some enterprises just don't sleep any more. Consequently, we do not just expect a rapid response; we are used to being able to contact suppliers when it suits us.

So, not only is customer care important, the minimum standard is becoming increasingly high. Do your staff and colleagues reach those standards? Maybe not if they share some of the common poor perceptions about customers.

Poor customer care perceptions

How well do your staff and subcontractors understand customer care?

We might know that customer care is vital to our business, but what do others think? Here are some of the common perceptions about customer care. See if you recognize any of them. If you do, I have also provided some information to help you counter them.

- *'Customers buy on price'*
  No one denies that price is important, but the majority of consumers see service as equally important, if not more so. And those consumers probably also work in the companies that are your customers, and some of them will be decision-makers there. If you always compete on price, it is easy for someone to better your price. If, however, you compete on service too, that is a much easier area in which to excel.

- *'Good products sell no matter what the level of service'*
  This is very unlikely to be true unless you have a genuine monopoly on a process or a patent on something everyone wants. (And even if this is the case, what about the cost of customer churn and loss of lifetime value?) The truth is that customers are attracted to good service. Good products need good service to back them up.

- *'It's OK to lose a customer as long as the loss is balanced by getting another one'*
  Not so. Attracting a new customer can cost as much as 15 times more than retaining an existing customer. Bad customer service can cost a typical medium-sized company

£1.8 billion in lost revenues and £267 million in lost profit. Reducing customer service problems by 1 per cent can increase profits of a typical medium-sized company by £16 million over five years. Eliminating customer service problems could double profit growth over five years. In some industries, reducing customer attrition by 5 per cent can double profits.

- *'As long as customers are satisfied overall, they'll stay with their current supplier'*
  Customers don't maintain a balance sheet of how they feel about a supplier; it tends to be the last interaction (good or bad) that determines how they feel about them. To be safe, therefore, you have to ensure that every interaction is a satisfying one and – and this is the worrying bit for an industry such as construction that uses many outsourced services – that includes interactions with anyone associated with the contract.

- *'Customers would rather have a product or service that does what it's supposed to do rather than lots of "Have a nice day" stuff'*
  But that still won't satisfy customers because, even if the product or service does what it's supposed to do, and yet customers have not enjoyed the transaction, they feel dissatisfied. What they'd rather have is a product or service that does what it's supposed to do and the level of care and concern appropriate to the transaction.

- *'There isn't much that I, as an individual, can do to enhance customer satisfaction'*
  Despite advances in modern technology, the 'human touch' is still at a premium during customer interactions. Whether it's resolving a problem for a customer, making an effort to see things from their point of view or just giving a genuinely warm greeting, you're doing your bit.

Think about the difference between your work and that of your competitors. If you're all technically OK with similar prices, the only difference is the service you provide.

You might find that, to enhance the level of customer care in your company, you have to gradually start educating people about customers. That will be easier if you understand some of its key concepts.

Key concepts in customer care

Like many subjects, customer care attracts its own terminology. Described below are some of the terms you will find in customer care, together with their benefits.

Human factor

The human factor is that part of a supplier–customer interaction relating to personal interaction.

As discussed earlier, when customers interact with a supplier their needs can be divided into basic needs and human needs. Basic needs relate to the essence of the transaction. Human needs relate to how they *feel* as a result of the transaction. This is the human factor.

The human factor is crucial to customer care because, despite advances in modern technology, the 'human touch' is still at a premium during customer interactions. As we have seen, service is overtaking price as the dominant criterion for many people and customer expectations are rising throughout all businesses and industry sectors.

So, whether you're resolving a problem for a customer, making an effort to see things from their point of view or just giving a genuinely warm greeting, you're doing your bit to capitalize on the human factor.

Moments of truth

A moment of truth is a short contact between supplier and customer during which the customer decides whether the company really is customer-oriented.

Whilst suppliers tend to see the supply of their product or service as a continuous process, customers do not. Customers see it as a series of 'snapshot' interactions. Each of those interactions is an opportunity for the customer to determine whether or not the supplier deserves their business.

Moments of truth are significant because customers' feelings towards a company are not an average of the feelings experienced throughout numerous contacts; they tend to be influenced by the last contact. A poor last contact, therefore, can wipe out the goodwill built up over previous contacts. By the same token, an exceptionally good last contact might make up for previous poor contacts. To be safe, however, you have to ensure that every interaction is a satisfying one. Here are some useful guidelines:

- Remember that, although your job might be described in terms of the functions you perform, customers' judgements are made in the light of their experience of you. So, see the main purpose of your job as *making customers glad they came into contact with you.*

- Remember that customers can come into contact with you directly (face-to-face, via the phone, post or e-mail) and indirectly by seeing something associated with you (such as your office, your staff, your behaviour towards others, your vehicle and so on).

- Think about your facial expression. Is it welcoming or offputting? Does it signal that you are there to help the customer or that you resent the interruption? Would you want to talk to someone who was wearing your expression? If in doubt, smile and make eye contact more deliberately than you do at the moment.

- When customers complain or are upset, remember that they are not getting at you personally. You just happen to be the company representative to whom they are talking. Avoid excuses and defensiveness (for example, saying, 'No one else has complained'). Instead, listen, then summarize, and then suggest what you can do to help them.

- When you can't do exactly what a customer wants, offer them more than one alternative. People prefer having a choice.

Service recovery

Problems arise in the best of companies. Resolving the situation to the customer's satisfaction is known as 'service recovery'. It is important for two main reasons. First, less than half the customers who experience a problem will tell the supplier about it. Not only will they find another supplier, but they will also tell as many as 16 other people about the poor service they've received.

Second, satisfying a dissatisfied customer usually turns them into an exceptionally loyal customer (often twice as loyal as customers who have never experienced a problem with you). Remember, reducing customer 'churn' increases 'lifetime value', which is a major contributor to company profitability.

Lifetime value

Lifetime value is the value of a customer (for example, how much money they spend or their profitability) for the time they remain a customer. The figures can be surprisingly high, as illustrated earlier.

Churn

Churn is the rate at which a company loses customers. It is a serious business issue because of the cost of replacing them, as illustrated earlier.

Strategic customer care

Are your staff and subcontractors your greatest asset or your biggest liability?

Although not everybody works in a position where they can win new business, it can be surprising how many people work in a position where they can lose business. (An example that is easy to relate to is the washing machine repair engineer who asks sarcastically, 'Who sold you this rubbish then?', thus guaranteeing that the customer never again buys another washing machine of that make or from that store.)

In construction, the culprit won't be a washing machine repair engineer. Instead it might be someone who is sorting out the snagging list or one of numerous subcontractors working in direct contact with your customers. The potential for lost business is real and points to the need for customer care to be part of your organization's culture and for your 'brand values' to percolate their way throughout the company, and even subcontractors, so that a consistent image is presented to customers.

Consequently, customer care is a strategic imperative for many organizations. It means interacting with customers in a way that makes them want to do, and continue doing, business with you. The focus of customer care has widened from just getting close to customers to include securing and growing their *lifetime value*. It may cost your company huge sums of money to acquire a customer. Profitability, therefore, relies on keeping that customer. Consequently, customer care is not just a nice thing to do; it is an effective way of saving money and adding to the bottom line.

Customer care is also one of the main ways in which organizations can differentiate their products or services from those of their competitors. With so many organizations offering similar products and services at similar prices through similar media, customers need a reason to do business with you rather than anyone else. That reason very often 'boils down' to how they *feel* as a result of doing business with you. At one extreme that can be affected by the quality of a senior manager's presentation and, at the other extreme, by how a junior clerk answers the telephone or the 'coffee stains' that are evident to customers.

The term 'coffee stain' derives from the fact that if airline passengers see coffee stains on the flip-down tables, they assume that the airline doesn't maintain the aircraft engines properly. The logic is, 'If they're this lax over the bits we can see, what are they like on the bits we can't see?'. It is a logic that travels widely. A mechanic's finger marks on your car might suggest that the service has not been carried out properly. A waiter's soiled clothes might suggest that the restaurant kitchen is unhygienic. Typing errors on a letter from your bank might suggest that they make errors in your account and so on. The common feature shared by all 'coffee stains' is that suppliers believe they are unimportant, yet they have the potential to trigger negative assumptions in customers' minds. Hence the importance of the impression given to customers.

Two facts can be added to this point. First, that impression can come from numerous sources: previous work customers see, your staff's technical ability, your written proposals, the way your staff answer the telephone, the attitude of your subcontractors and so on. Second, it can take just one small incident to undo a good impression. Combine all these points and there is really no viable alternative to making customer care part of your organization's culture. The easiest way to do that is to recognise the importance of the supplier–customer chain.

The supplier–customer chain

The level of service one person provides a customer is often dependent on the level of service someone has provided them – this is the supplier–customer chain. This means that if people within your organization who may have no contact with real customers improve the service they provide their colleagues, it will be easier for those colleagues to satisfy real customers. When customer care becomes part of an organization's culture in this way, it can lead to greater internal efficiency, teamwork and job satisfaction.

The concept of the supplier–customer chain extends even further with the concept of the *value chain*.

The value chain

Originally *value chain analysis* was an analytical accounting technique intended to identify the profitability of separate steps in complex manufacturing processes to determine where cost improvements could be made and value creation improved.

The *value chain* concept has now been extended to include the entire production process of a product or service from beginning to end regardless (and this is the important bit) of who owns any particular value-adding step. In addition to people within your organization, therefore, it includes your suppliers, their suppliers, your customers and their customers.

Value chain management is an operating strategy that emphasizes linking your organization with others in the same value chain to create mutually beneficial relationships that, in turn, will create greater success than the individual enterprises could achieve in isolation. You might do so, for example, by seeing how you can assist one another in improving quality or customer service throughout the entire value chain to improve sales to the final customer. You might even extend it to sharing planning, inventory, HR and IT systems, corporate cultures and so on. You might also agree how each of you in the chain might help better serve each other's customer.

Value chain management is particularly important in the construction industry which is often typified by two significant features: first, its extensive use of subcontractors; and, second, the large size of many construction projects, which makes it possible for a customer to employ several specialist construction companies working together. Far too often the result is less satisfactory than it could be because one 'partner' feels that another is guilty of delays and snags that cause problems with the customer.

Ensure your subcontractors deliver as good a level of service as you do yourself by treating them as you would like them to treat your customers.

Even small construction contracts tend to involve numerous suppliers and subcontractors specializing in different trades. The larger contracts require coordination between even more suppliers and subcontractors, and the more links in the chain, the greater the risk of there being weak links. When a weak link becomes evident, a great deal of time, energy and, therefore, money gets diverted from getting the job done to the detriment of customer care, with all the accompanying negative effects on winning new business from that customer. Weak links, therefore, need to be minimized, and one way of doing that is to apply the value chain concept by:

- establishing long-term relationships with trusted suppliers and subcontractors

- agreeing procedures and efficient ways in which you can help each other

- communicating the significance of good relations with others in the value chain

- ensuring maximum openness and cooperation between suppliers and customers

- treating your own suppliers and subcontractors with the same customer care that you wish them to apply to you and your customers.

Customer relationship management

Customer relationship management (CRM) is a process of managing relationships with customers to increase the benefits of that relationship to both you and them. Basically, it involves knowing about your customers in sufficient detail so that you can tailor the way in which you interact with them to create a mutually beneficial relationship. Its simple purpose is to increase their lifetime value.

Although, as a term, CRM has only been around since the 1980s, as an approach to customers it's been around for hundreds of years. The most obvious example is a small corner shop run by the proprietor, who knows all the customers personally, knows their preferences, what they usually buy, how they like to be treated and so on. In bigger companies, however, salespeople may no longer know their customers personally and that degree of detailed knowledge is lost. CRM uses modern technology to help put this element back so that, say, an order clerk can see the nature of the company's relationship with that customer from their computer screen.

Furthermore, computer technology bestows yet another benefit that the corner shop proprietor could only dream about in that it allows companies to pinpoint different markets amongst their customers. So, if a supermarket wants to promote a new product to young, sports-loving parents in a certain socioeconomic stratum within a certain geographical area it has the technology to do so. It's these two abilities combined (customer knowledge and pinpoint marketing) that gives CRM its massive potential in many industries. It can lead to greater customer satisfaction, less customer churn and more cross-selling. But how can it be used to win more business in construction?

As in many industries, some construction industry customers make one-off, low-profit purchases. It is rarely worth spending huge sums of money attempting to build up relationships with them (unless you can convert them into the sorts of customers who make more frequent purchases or higher-value purchases). On the other hand, the lifetime value of customers who make repeated, high-profit purchases could be so huge that it would definitely be worth investing time and effort building a solid relationship with them.

How can construction industry companies build solid relationships? Well, building on the essential foundation of technical excellence, value for money and great customer care, consider some of the ideas from Chapter 1 on marketing.

- Create a list of the benefits of working with you and, by checking this list against each customer, identify any benefits a customer is not receiving. Then find a way of offering that benefit to them.
- Use information gained when networking and probing (using the persuasive funnel) to identify each customer's

potential problems and propose how you might help solve them.

- Write or circulate articles or factsheets relevant to target customers.

- Write and circulate newsletters containing information relevant to targeted customers.

- Run seminars on topics relevant to targeted customers.

- Organize and facilitate a 'customer club' bringing together like-minded customers and helping them network and learn from one another.

- Engage in complementary marketing for customer needs that are not in your repertoire.

- Create a told/sold matrix for your customers (see Figure 6.1).

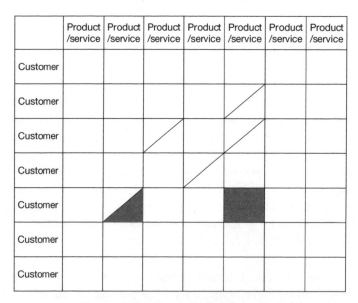

	Product /service	Product /service	Product /service	Product /service	Product /service	Product /service	Product /service
Customer							
Customer							
Customer							
Customer							
Customer							
Customer							
Customer							

Figure 6.1   *The told/sold matrix*

A told/sold matrix avoids that uncomfortable situation in which you find that a good customer has bought from a competitor something you could have easily supplied yourself because 'We didn't know you did that too'. On one axis of your matrix you list your customers and on the other axis you list everything you can supply. Where a customer is already using a product or service, blank out the appropriate square. Where they are not using, but could use, a product or service, put a diagonal line

Do your customers know everything from which they could benefit?

through the square. When you have told them about the product or service, blank out half the square. When they have bought it, you blank out the remaining half of the square. It is a simple and effective way of winning new business from existing customers.

- Create a CRM matrix and decide, based on your current relationship with each customer, in which quadrant each customer would naturally fit.

Transaction frequency

Large, one-off, high-profit purchases	**Large, ongoing, high-profit purchases**
• Identify high value customers. • Pay special attention to them. • Deliberately differentiate the service they receive so that they begin to desire more personal attention and so encourage more loyalty from them. • Maximize their lifetime value.	• Interact with them. • Identify, and cater for, their individual preferences. • Turn interaction into a beneficial experience for them. • Work as 'partners' with them.
Small, one-off, low-profit purchases	**Small, ongoing, low-profit purchases**
• Deal with customers swiftly and efficiently so that they prefer dealing with you/your brand than anyone else.	• Deal with customers swiftly and efficiently, but also identify them personally.

(left axis: **Transaction value**)

Figure 6.2 *The CRM matrix*

An essential part of CRM is recognizing the types of transaction you have with customers, and the CRM matrix helps you do that. It has two axes – one for transaction value and one for transaction frequency – which, in turn, provide four quadrants. The quadrant a customer is in will affect the way a supplier interacts with them. For example, an airline passenger travelling first-class to New York expects a different interaction with the airline staff than a passenger travelling across town by bus. Although the bus company will naturally seek to retain the bus passenger, it will do so by being punctual and reliable rather than through anything more personal because the low transaction value makes anything else unprofitable. The airline, however, will try to ensure the first-class passenger's loyalty not only with reliability and punctuality, but also with a special first-class lounge, a special magazine published several times a year, recorded personal preferences that staff can act on when booking and during the flight, preferential discounts and so on, because the high transaction value makes all this financially worthwhile.

Knowing in which quadrant your customer fits enables you to take two business-winning actions. First, you can tailor your interaction with customers to retain them longer. Second, you can target certain customers with specific marketing activity to help them graduate from one quadrant to another (by increasing their order value, their order frequency or both).

Action points

1 Conduct a customer care audit. Can you describe observable proof that your organization genuinely cares for customers?

2 Ensure that customer care features in all selection interviews, induction training and promotion decisions.

3 Audit company procedures and senior management decisions for inadvertent negative effects on customer care.

4 Audit employee care policies. (Employee care leads to customer care.)

5 Calculate the cost of customer churn and the financial benefits of improving it. Develop a plan accordingly.

6 Calculate typical lifetime customer value and the benefits of improving it. Develop a plan accordingly.

7 Review your SWOT analysis in the light of the customer trends described in this chapter.

8 As things can go wrong even in well-run companies, do
   you have a 'recovery' process? Are staff empowered to use
   it? What would be the attitude of senior management if a
   staff member 'overstepped the mark' but succeeded in
   retaining the customer's business?

9 Identify the moments of truth in your business processes
   and ensure that they are 'watertight'.

10 Introduce the value chain concept at your next team
   meeting or team development day and agree how you can
   implement it.

11 Introduce the CRM matrix at your next team meeting or
   team development day and, using it as an exercise, agree
   how you can implement it.

12 Explain the told/sold matrix to a new manager and, as part
   of their induction, get them to complete it with recommendations for at least 20 customers.

Chapter 7

Partnering

The subject of partnering is so important to the construction industry that it deserves its own chapter.

Definition

Partnering is a term used to describe *the practice of agreeing mutually beneficial, formal working relationships between organizations that work together.*

You might wonder what is so important about partnering that it deserves its own chapter; after all, organizations in every industry work with other organizations – they could not get their jobs done otherwise! Motor car manufacturers, for example, work with numerous suppliers and dealers. In construction, companies have always used subcontractors and specialists such as architects and surveyors. So what is different about 'partnering'? Let's begin by looking at the phrase 'mutually beneficial'.

Partnership and mutual benefit

For a long time, the construction industry has been typified by contractors being commercially 'tough' with their sub-contractors, screwing down prices and adopting an adversarial response to problems. While, thankfully, this is not always the case, it is sufficiently widespread to have been a feature of the construction industry since at least the mid-1930s when an architect called Alfred Bosson made the point that frag-mentation and adversarial attitudes created unnecessary costs which fed into overheads, seriously damaging competitiveness. He estimated the figure to be 15 per cent which is not far off the 10 per cent figure quoted by Sir Michael Latham 60 years later. Other more recent evidence has suggested that the true cost of fragmentation and adversarial attitudes is between 30 and 50 per cent. In other words, over the last 70 years the level of unnecessary costs caused by fragmentation and adversarial attitudes has, at best, remained unchanged and, at worst, has increased.

How much is it costing you to be 'tough' to work for? How much could you save if you worked in partnership?

And that is just the obvious cost. If you add the cost of wasted time, energy and stress that could be directed towards more profitable pursuits, the scale of the problem becomes alarming.

That is why, when Sir John Egan looked at reversing the trend and turning unnecessary costs into enhanced profits for both the construction industry and its clients, he must have recognized some of the features from other British industries. Many other industries had been typified by aggressive attitudes towards subcontractors but, to meet the challenges of global competition, they had no choice but to reinvent themselves. One aspect of that reinvention, and an important one, was partnering – a simple, written agreement defining the nature and quality of the relationship between the parties.

That agreement commits all parties to improved efficiency, cost reduction, quality improvements, openness and transparency, reasonable profits, shared risks and common goals. It typically involves integrating the design and construction process and incorporating the value chain management techniques now widespread in other industries (see Chapter 6, 'Winning New Business Through Customer Care'). It is, therefore, more than an informal relationship. It enables the partners to reap the significant benefits of cost and time reduction of working together – that is, *with* one another rather than *against* or *despite* one another.

Why bother?

There are three reasons for taking the trouble to build partnerships. First, the cost of not working collaboratively is immense. Not only could your overheads be significantly higher than your partnering competitors (or your profit margins significantly lower), but your job satisfaction and that of your staff will also be lower. (After all, you and your staff went into construction to build things not to argue with subcontractors!) Good staff want to work for good companies; poor companies have to make do with what is left.

Second, not only do you benefit from lower costs, higher profit, better staff and more job satisfaction, but you also benefit from the pooled wisdom, knowledge and creativity of your partners. This is the synergy effect that happens when people work *across* organizational boundaries to achieve a common goal. Quite simply, it takes longer, costs more and generates more hassle to produce poor-quality products than it does to produce high-quality work first time. Benefits being reported by construction companies that have embraced partnering include:

- empowerment of the architects to deliver quality

- exceptional functionality of the product

- an end to adversarial relationships

- a 'right first time' culture

- the release of the knowledge, experience and design skills of specialist suppliers

- up to 60 per cent reduction in labour and material costs

- over 100 per cent increase in productivity

- 18–25 per cent reduction in construction time

- 10–14 per cent reduction in the cost of ownership

- the use of high-quality, long-life materials and components, to bring a consequent reduction in operating expenditure

- fewer accidents

- better training.

As one housing association executive put it:

'We have customer satisfaction levels above 90 per cent, suppliers confident to invest in capital equipment.'

> *It is precisely those kinds of benefits that led us to conclude that we needed to introduce partnering to our procurement processes within the Housing Executive. I and my colleagues were fed up to the back teeth with adversarial attitudes, unnecessary recourse to litigation (and the associated cost burden), unrealistically low tenders (almost as bad as unrealistically high), crap workmanship, low tenant satisfaction, low tenant expectations, chasing bad workmanship (and the associated cost burden) rather than co-operating to achieve quality workmanship. Already, we have customer satisfaction levels above 90%, suppliers confident to invest in capital equipment, willing to recruit apprentices due to work flow security, greater job satisfaction etc.*

Third, the time is right. There is a substantial and growing programme of public-sector projects, renewing outdated schools, social housing and infrastructure. The growth of public–private partnerships and other procurement routes leaves UK firms well placed to win new business through partnering.

Making partnerships work

So what are the keys for successful partnerships? They are as follows:

- Ensure that mutual objectives are agreed at the beginning and reviewed throughout the partnership.

- Ensure a single reference point so that everyone knows that you are all signed up to the same goals and working as a single unit to achieve them.

- While guaranteeing confidentiality, ensure openness and transparency.

- Agree a systematic approach to resolving the inevitable problems – while they are still problems and not disputes. Ensure that problem resolution is solution-focused rather than blame focused. Use common sense more than contracts to resolve problems. To avoid escalating problems, resolve them at the lowest level possible. Accept that adversarial approaches waste time, money, energy and goodwill.

- Guarantee equality of rights between partners.

- Communicate more while reducing paperwork.

- Focus on value rather than cost.

- Help each other achieve best practice.

Food for thought

People in partnering arrangements think and act differently to people in traditional working relationships. Mutual self-interest, openness and trust, aligned objectives, a shared culture, equitable risk- and reward-sharing are the foundations of successful partnerships. To work this way, people need to see an example set by their managers and by senior managers and directors. A partnership is not just a piece of paper, it is a way of life. But it is a way of life with a practical purpose; you will win more business and the business you win will be more profitable.

Action points

1 Ask your subcontractors (especially those whose work you value) to describe how they benefit from working with you. If all you are to them is a source of work, think about partnering.

2 Estimate how much time and money your company spends on resolving problems and how much would be added to the bottom line if those problems did not arise or were resolved quickly and amicably at the lowest possible level.

3 Consider who you work with regularly (suppliers, specialists, subcontractors and so on) and consider how you could all benefit if your working relationship was more open, transparent and mutually helpful.

4 Find some companies from other industries or some non-competing companies from your industry that are into partnering and ask them about their experiences and what advice they would have for you.

5 Look at the benefits of partnering listed in this chapter and identify three that would have a major effect on your business. Plan how you will go about achieving them.

6 Take your management team to an off-site meeting. Introduce them to the concept and benefits of partnering and ask them each to agree personal goals to implement the partnering ethos in the way they work with other organizations and with their own staff.

Now that you have reached the end of this book, you probably have hundreds of profitable ideas swimming around your head. It is a good time now, therefore, to remind yourself of some essential points.

You went into construction to enjoy yourself and to make money, either by growing your own business or by being an effective employee. You will only achieve those aims if you are successful at turning inputs (materials and skills) into profitable outputs (the things your customers pay for). To do that, you have to be good technically. You also have to be good at winning new business.

This book takes a broad approach to winning new business. Whilst some people have roles that dedicate them to that purpose, in successful companies everyone contributes to the process. To help you, this book has given you insights and ideas relating to marketing, selling, negotiating, customer care and partnering. It has also suggested actions you can take that will help you put what you have learned into practice. This is a lot to absorb in one go. So keep *Winning New Business in Construction* handy. Encourage others to read it. See it as a source of reference, good ideas and stimulation. Above all, return to it regularly and frequently. After all, winning new business is a journey, not a destination. It is not something you learn how to do; it is something you keep getting better at. Hopefully, this book will help you.

Good luck.

Index

Project Management in Construction
Dennis Lock
Hardback   208 pages   244 x 172 mm   0 566 08612 3

Quality Management in Construction
Brian Thorpe and Peter Sumner
Hardback   234 pages   244 x 172 mm   0 566 08614 X

Improving People Performance in Construction
David Cooper
Hardback   184 pages   244 x 172 mm   0 566 08617 4

Health and Safety in Construction Design
Brian Thorpe
Hardback   112 pages   244 x 172 mm   0 566 08615 8

For more details of these books or new titles in the series visit www.gowerpub.com or contact our sales department:
Sales Department, Gower Publishing Limited, Gower House, Croft Road, Aldershot, Hants, GU11 3HR, UK. Tel: +44 (0)1252 331551, e-mail: info@gowerpub.com

Join our e-mail newsletter

Gower is widely recognized as one of the world's leading publishers on management and business practice. Its programmes range from 1000-page handbooks through practical manuals to popular paperbacks. These cover all the main functions of management: human resource development, sales and marketing, project management, finance, etc. Gower also produces training videos and activities manuals on a wide range of management skills.

As our list is constantly developing you may find it difficult to keep abreast of new titles. With this in mind we offer a free e-mail news service, approximately once every two months, which provides a brief overview of the most recent titles and links into our catalogue, should you wish to read more or see sample pages.

To sign up to this service, send your request via e-mail to info@gowerpub.com. Please put your e-mail address in the body of the email as confirmation of your agreement to receive information in this way.

GOWER